Lecture Notes in
Operations Research and
Mathematical Systems
Economics, Computer Science, Information and Control

Edited by M. Beckmann, Providence and H. P. Künzi, Zürich

55

P. A. V. B. Swamy

Statistical Inference in Random Coefficient Regression Models

Springer-Verlag
Berlin · Heidelberg · New York

This series aims to report new developments in mathematical economics and operations research and teaching quickly, informally and at a high level. The type of material considered for publication includes:

1. Preliminary drafts of original papers and monographs
2. Lectures on a new field, or presenting a new angle on a classical field
3. Seminar work-outs
4. Reports of meetings

Texts which are out of print but still in demand may also be considered if they fall within these categories.

The timeliness of a manuscript is more important than its form, which may be unfinished or tentative. Thus, in some instances, proofs may be merely outlined and results presented which have been or will later be published elsewhere.

Publication of *Lecture Notes* is intended as a service to the international mathematical community, in that a commercial publisher, Springer-Verlag, can offer a wider distribution to documents which would otherwise have a restricted readership. Once published and copyrighted, they can be documented in the scientific literature.

Manuscripts

Manuscripts are reproduced by a photographic process; they must therefore be typed with extreme care. Symbols not on the typewriter should be inserted by hand in indelible black ink. Corrections to the typescript should be made by sticking the amended text over the old one, or by obliterating errors with white correcting fluid. Should the text, or any part of it, have to be retyped, the author will be reimbursed upon publication of the volume. Authors receive 75 free copies.

The typescript is reduced slightly in size during reproduction; best results will not be obtained unless the text on any one page is kept within the overall limit of 18 x 26.5 cm (7 x 10 ½ inches). The publishers will be pleased to supply on request special stationery with the typing area outlined.

Manuscripts in English, German or French should be sent to Prof. Dr. M. Beckmann, Department of Economics, Brown University, Providence, Rhode Island 02912/USA or Prof. Dr. H. P. Künzi, Institut für Operations Research und elektronische Datenverarbeitung der Universität Zürich, Sumatrastraße 30, 8006 Zürich.

Die „*Lecture Notes*" sollen rasch und informell, aber auf hohem Niveau, über neue Entwicklungen der mathematischen Ökonometrie und Unternehmensforschung berichten, wobei insbesondere auch Berichte und Darstellungen der für die praktische Anwendung interessanten Methoden erwünscht sind. Zur Veröffentlichung kommen:

1. Vorläufige Fassungen von Originalarbeiten und Monographien.
2. Spezielle Vorlesungen über ein neues Gebiet oder ein klassisches Gebiet in neuer Betrachtungsweise.
3. Seminarausarbeitungen.
4. Vorträge von Tagungen.

Ferner kommen auch ältere vergriffene spezielle Vorlesungen, Seminare und Berichte in Frage, wenn nach ihnen eine anhaltende Nachfrage besteht.

Die Beiträge dürfen im Interesse einer größeren Aktualität durchaus den Charakter des Unfertigen und Vorläufigen haben. Sie brauchen Beweise unter Umständen nur zu skizzieren und dürfen auch Ergebnisse enthalten, die in ähnlicher Form schon erschienen sind oder später erscheinen sollen.

Die Herausgabe der „*Lecture Notes*" Serie durch den Springer-Verlag stellt eine Dienstleistung an die mathematischen Institute dar, indem der Springer-Verlag für ausreichende Lagerhaltung sorgt und einen großen internationalen Kreis von Interessenten erfassen kann. Durch Anzeigen in Fachzeitschriften, Aufnahme in Kataloge und durch Anmeldung zum Copyright sowie durch die Versendung von Besprechungsexemplaren wird eine lückenlose Dokumentation in den wissenschaftlichen Bibliotheken ermöglicht.

Lecture Notes in
Operations Research and
Mathematical Systems

Economics, Computer Science, Information and Control

Edited by M. Beckmann, Providence and H. P. Künzi, Zürich

55

P. A. V. B. Swamy

Ohio State University, Department of Economics,
Columbus, Ohio/USA

Statistical Inference in Random Coefficient Regression Models

Springer-Verlag
Berlin · Heidelberg · New York 1971

AMS Subject Classifications (1970): 62 J 05, 62 M 10, 62 P 20, 90 A 20

ISBN 3-540-05603-3 Springer-Verlag Berlin Heidelberg New York
ISBN 0-387-05603-3 Springer-Verlag New York Heidelberg Berlin

Offsetdruck: Julius Beltz, Hemsbach/Bergstr.

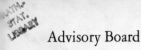

To my parents and sister

PREFACE

This short monograph which presents a unified treatment of the theory of estimating an economic relationship from a time series of cross-sections, is based on my Ph.D. dissertation submitted to the University of Wisconsin, Madison. To the material developed for that purpose, I have added the substance of two subsequent papers: "Efficient methods of estimating a regression equation with equi-correlated disturbances", and "The exact finite sample properties of estimators of coefficients in error components regression models" (with Arora) which form the basis for Chapters II and III respectively. One way of increasing the amount of statistical information is to assemble the cross-sections of successive years. To analyze such a body of data the traditional linear regression model is not appropriate and we have to introduce some additional complications and assumptions due to the heterogeneity of behavior among individuals. These complications have been discussed in this monograph. Limitations of economic data, particularly their non-experimental nature, do not permit us to know <u>a priori</u> the correct specification of a model. I have considered several different sets of assumptions about the stability of coefficients and error variances across individuals and developed appropriate inference procedures. I have considered only those sets of assumptions which lead to operational procedures. Following the suggestions of Kuh, Klein and Zellner, I have adopted the linear regression models with some or all of their coefficients varying randomly across individuals. The random coefficient approach considered in this study provides a simple and elegant procedure for handling inter-individual heterogeneity in a cross-section. The traditional fixed coefficient models can be treated as special cases of random coefficient models.

It is presumed that the reader has a reasonable familiarity with matrix notation and operations, with concepts of quadratic forms and statistical estimation. My thesis on random coefficient regression models grew from Zellner's article: "On the aggregation problem: A new approach to a troublesome problem" and Rao's article: "The theory of least squares when the parameters are stochastic and its application to the analysis of growth curves".

I owe a debt of gratitude to Professor A. S. Goldberger, my thesis advisor, for his criticism and helpful suggestions at various stages of my work. His searching criticism has been of great value in shaping my thoughts on this problem. Despite his busy work he carefully went through my thesis and improved my presentation. He also helped me in getting free computer time. I am indebted to Professor A. Zellner who offered me Research Assistantship for two years from 1964 to 1966. During this period, though I did not do any research on my own, my association with him gave me deeper understanding of economic behavior and the problems of incorporating them in model construction and econometric analysis. He was kind enough to comment on various parts of my thesis. In Summer 1968 he offered me Research Associateship and extended an excellent opportunity to prepare my thesis material for publication. It was entirely due to his efforts Springer Verlag, New York, agreed to publish my thesis. My gratitude goes to him. Responsibility of any possible remaining errors is mine.

My thanks also go to Mr. Richard D. Porter for programming assistance. For expert typing of difficult material I am grateful indeed to Mrs. Shirley Black of the Graduate School of Business, University of Chicago and to Mrs. Lily Shultz.

My work in this thesis is financed by National Science Foundation under Grants GS-151 and GS-1350.

Finally, I express my sincere thanks to Professor J.N.K. Rao and Dr. T. R. Rao who were instrumental in bringing me to the United States -- a venture whose main product is this thesis.

Columbus, Ohio
June 1971.

P.A.V.B. Swamy

TABLE OF CONTENTS

CHAPTER I

INTRODUCTION

1.1 Purpose and Outline of the Study

The purpose of this study is to examine various approaches which are useful in estimating a relation from the combined time series and cross-section data. The study in this thesis can be divided into two parts. In the first part following Kuh (1959, 1963) we consider a regression equation with additional effects associated with both the time direction and cross-sectional units and develop efficient methods of estimating the parameters of that model under different sets of assumptions. In the second part we examine critically the specifying assumptions traditionally entertained while analyzing cross-section behavior. In doing so, we will consider Klein's (1953, pp. 216-218) suggestion to treat the coefficients of a regression model as random in cross-section analyses to account for spacial or interindividual heterogeneity. Under alternative sets of assumptions, we will study random coefficient regression (RCR) models which treat both intercept and slope coefficients as random variables and will develop appropriate statistical inference procedures. Following Zellner (1966), we discuss the implications of RCR models for the aggregation problem. Further, we discuss the problems of identification, forecasting, and of using stochastic prior information within the framework of RCR models.

For empirical implementation of RCR models considered in this thesis, panel data provide the base. By panel data we mean temporal cross-section data obtained by assembling cross-sections of several years, with the same cross-section units appearing in all years. The cross-section units could be households, firms or any micro economic units. Data on a number of units for a single year with one observation per unit constitute a single cross-section. We apply our statistical procedures in estimating the parameters of a random coefficient investment model using panel data on certain corporations in the U.S.A. We also use aggregate time series data on each of twenty-four countries to estimate the aggregate consumption function with coefficients varying randomly across countries.

The plan of the thesis is as follows: We devote the remaining sections of this chapter to a review of the literature, to an explicit statement of the assumptions which are appropriate for a cross-section or a temporal cross-section analysis, and to an indication of the statistical problems connected with such analyses. In Chapter II we discuss efficient methods of estimating a regression equation with equicorrelated disturbances. In Chapter III we develop efficient methods of estimating the parameters of an equation with additional random-effects associated with both the time direction and cross-sectional units. In Chapter IV we specify a linear regression model with random coefficients and develop the relevant statistical inference and forecasting procedures. We apply these procedures to analyze certain investment and consumption functions in Chapters V and VI, respectively. We consider the problems of identification and incorporating prior information in Chapter VII.

1.2 Review of the Literature on Regression Models with Random and Fixed Coefficients

(a) Pooling Cross-Section and Time Series Data in the Estimation of a Regression Model

The discussion in this section is based on Kuh (1959), Mundlak (1961), Nerlove (1965, pp. 157-187), and Balestra and Nerlove (1966). Nerlove considered the following dynamic model to analyze a time series of cross-sections.

$$y_{it} = \beta_0 + \sum_{k=1}^{K-1} \delta_k z_{kit} + \sum_{g=1}^{G} \gamma_g y_{it-g} + u_{it}$$

$$(i = 1,2,\ldots,n \;;\; t = 1,2,\ldots, T) \quad . \qquad (1.2.1)$$

We have observations on y's, and z's for n individual units. The disturbance term u_{it} is decomposed into two statistically independent parts: an individual effect, and a remainder. Thus,

$$u_{it} = \mu_i + \nu_{it} \quad . \qquad (1.2.2)$$

Nerlove assumes that the random variables μ_i and ν_{it} are independent such that

$$E\mu_i = E\upsilon_{it} = 0 \text{ for all } i \text{ and } t \quad ; \tag{1.2.3a}$$

$$E\mu_i\upsilon_{it} = 0 \text{ for all } i \text{ and } t \quad ; \tag{1.2.3b}$$

$$E\upsilon_{it}\upsilon_{i't'} = \begin{cases} \sigma_\upsilon^2 & \text{if } i=i' \text{ and } t=t' \ , \\ 0 & \text{otherwise;} \end{cases} \tag{1.2.3c}$$

$$E\mu_i\mu_{i'} = \begin{cases} \sigma_\mu^2 & \text{if } i=i' \ , \\ 0 & \text{otherwise .} \end{cases} \tag{1.2.3d}$$

He further assumes that the υ_{it} and the μ_{it} are independent of themselves for different individuals.

In his important contribution on empirical production function free of management bias, Mundlak (1961) specifies a log linear production function with nonstochastic regressors and includes a time invariant individual firm effect μ_i. He treats the μ_i's as fixed parameters obeying the restriction $\sum_{i=1}^{n} \mu_i = 0$. In this case the coefficients can be estimated by applying ordinary least squares (OLS) to observations expressed as deviations from their respective individual firm means. Kuh (1959) also specifies a linear regression model with nonstochastic regressors and includes an individual time invariant random effect μ_i. Unlike Nerlove, Kuh assumes that μ_i is correlated with υ_{it}.

If we combine (1.2.1) and (1.2.2) and write

$$y_{it} = (\beta_0 + \mu_i) + \sum_{k=1}^{K-1} \delta_k z_{kit} + \sum_{g=1}^{G} \gamma_g y_{it-g} + \upsilon_{it} \quad , \tag{1.2.4}$$

it is clear that the dynamic model as described in (1.2.1)-(1.2.3d) has an intercept varying randomly across units and fixed slopes. Under the assumptions (1.2.3a)-(1.2.3d) the variance-covariance matrix of u_{it}'s is a block diagonal matrix and Aitken's (1934-35) generalized least squares estimator of coefficients in (1.2.1) is a best linear unbiased estimator (BLUE). The Aitken estimator is not operational if the variances of error terms are unknown.

Nerlove presented an interesting discussion on the application of the maximum likelihood (ML) method to estimate the parameters in the model (1.2.1). He concluded that the ML method was inapplicable to the above situation if it produced an estimate of $\rho = \sigma_\mu^2/(\sigma_\upsilon^2 + \sigma_\mu^2)$, outside the unit interval [0,1) which was closed on the left and open on the right. He suggested an alternative two-step procedure which was similar to two-stage least squares, to get consistent estimates of the parameters. This method, however, breaks down when there are no truly exogenous variables entering into the model with nonzero coefficients. Therefore, he suggested the following alternative procedure.

Since the υ_{it}'s are serially independent, provided sufficient number of observations over time is available, the OLS estimates of parameters in (1.2.4) for each individual may be found and they are consistent. We get n sets of estimates for γ's and δ's. Let these be denoted by $\hat{\delta}_{ki}$ $(i = 1,\ldots,n\ ;\ k = 1,\ldots, K - 1)$ and $\hat{\gamma}_{gi}$ $(i = 1,\ldots, n\ ;\ g = 1,\ldots, G)$. The subscript i indicates the unit to which the estimate refers. $\beta_0 + \mu_i$ differs from one individual to another. By assumption, the parameters $\delta_1,\ldots,\delta_{K-1}$ and $\gamma_1,\ldots,\gamma_G$ are the same for all individuals. Consequently, the above estimates are combined as

$$\hat{\delta}_k = \frac{1}{n}\sum_{i=1}^{n} \hat{\delta}_{ki} \qquad (k = 1,2,\ldots,K - 1) \qquad (1.2.5a)$$

$$\hat{\gamma}_g = \frac{1}{n}\sum_{i=1}^{n} \hat{\gamma}_{gi} \qquad (g = 1,2,\ldots,G) \qquad (1.2.5b)$$

$$\hat{\beta}_0 = \frac{1}{n}\sum_{i=1}^{n} \left\{\bar{y}_i - \sum_{k=1}^{K-1} \hat{\delta}_k \bar{z}_{ik} - \sum_{g=1}^{G} \hat{\gamma}_g \bar{y}_{i,-g}\right\} \qquad (1.2.5c)$$

where $\bar{y}_{i,-g} = \frac{1}{T}\sum_{t=1}^{T} y_{it-g}$ $(g = 0,1,\ldots,G)$; $\bar{z}_{ik} = \frac{1}{T}\sum_{t=1}^{T} z_{kit}$. The variances are estimated by

$$\hat{\sigma}_\mu^2 = \frac{1}{n}\sum_{i=1}^{n} \left[\left(\bar{y}_i - \sum_{k=1}^{K-1} \hat{\delta}_k \bar{z}_{ik} - \sum_{g=1}^{G} \hat{\gamma}_g \bar{y}_{i,-g}\right) - \hat{\beta}_0\right]^2 , \qquad (1.2.5d)$$

and

$$\hat{\sigma}^2 = \frac{1}{nT}\sum_{i=1}^{n}\sum_{t=1}^{T} \left(y_{it} - \hat{\beta}_0 - \sum_{k=1}^{K-1} \hat{\delta}_k z_{kit} - \sum_{g=1}^{G} \hat{\gamma}_g y_{it-g}\right)^2 , \qquad (1.2.5e)$$

where $\sigma^2 = \sigma_\mu^2 + \sigma_\upsilon^2$.

The estimates of σ_μ^2 and σ^2 are substituted in the Aitken's generalized least squares estimator in the second round, to get asymptotically more efficient estimates of the parameters. This is not the same as that finally chosen in the gas study by Balestra and Nerlove (1966) but does bear a strong resemblance to the two-round instrumental variable procedure adopted in the gas study. We will show in Chapter II that there exists a whole class of consistent and asymptotically efficient estimators of coefficients in (1.2.4) and they can be adopted even when $T < (K+G)$. We will study their small sample properties when all the independent variables are nonstochastic and indicate different estimators which are efficient for different parameter values and sample sizes.

(b) Another Specification in Temporal Cross-Section Analyses

In order to account for the effects associated with both the time direction and cross-sectional units, Hoch (1962), Mundlak (1963) and Kuh (1963) have considered the following relation

$$y_{it} = \beta_0 + \delta_1 x_{1it} + \dots + \delta_{K-1} x_{K-1it} + \mu_i + \tau_t + \nu_{it} \quad , \qquad (1.2.6)$$

where y_{it} is the dependent variable for the ith individual in the tth year, $x_{kit}(k = 1,2,\dots,K-1)$ are independent variables, the term μ_i is a constant through time; it is an attribute of the ith individual which is unaccounted for by the included cross-section variables but differs among individuals, the term τ_t is the same for all individuals at a given time but varies through time, and the term ν_{it} differs among individuals both at a point in time and through time. The μ_i, τ_t and ν_{it} are unobserved variables.

If the μ_i's and τ_t's are fixed parameters obeying the restrictions $\sum_{i=1}^{n} \mu_i = 0$ and $\sum_{t=1}^{T} \tau_t = 0$, respectively; and the ν_{it}'s are serially uncorrelated and independent from one individual to another with zero mean and a common variance, then the analysis of covariance methods can be adopted in estimating the parameters appearing in (1.2.6), cf. Graybill (1961, pp. 396-403).

Mundlak (1963) and Hoch (1962) in their work on the estimation of production and associated behavioral functions from a combined cross-section and time series data adopted the analysis of covariance methods.

If the μ_i's, τ_t's and ν_{it}'s are random variables with zero means and certain unknown variances, then the variance-covariance matrix of the error vector in (1.2.6) will be nonscalar and unknown. We may have to consider an Aitken procedure based on the estimated variances and covariances of error terms if we want to obtain estimates of coefficients with desirable properties, cf. Wallace and Hussain (1969) and Hussain (1969). We discuss this procedure further in Chapter III and throw some light on the small sample properties of different coefficient estimators.

It is easy to recognize that the model (1.2.6) becomes a mixed model with random intercept and fixed slopes when the μ_i's and τ_t's are random variables. When we use a single cross-section data for a given year t_0, $\mu_i + \nu_{it_0}$ is a random element in the intercept term. Assuming that the regression coefficient vector is the same for all individuals, if we use aggregate time series data to estimate an equation of the type (1.2.6), τ_t is the random element in the intercept term since the terms μ_i and ν_{it} tend to average out to zero when aggregated over all units.

(c) Assumptions Appropriate to the Analysis of Cross-Section Behavior

Most of the empirical work concerned with estimation of a behavioral equation using cross-section data has been based on the assumption that the equation is valid for each of the different individual units in a sample, and adequately describes the mechanism generating the data on certain economic features of those units. In some aggregate studies the coefficient estimates are interpreted as if all the micro units which make up the aggregate have the same regression coefficient vector.

The nature of these assumptions will be clear if we read Klein's (1953, pp. 211-225) brief but illuminating discussion on estimation from cross-section data. He uses the following simple linear relation for illustrative purposes.

$$y_{it} = \beta_0 + \beta_1 x_{it} + u_{it} \quad , \tag{1.2.7}$$

where y_{it} is the dependent variable for individual i at time t, x_{it} is the independent variable for individual i at time t, and u_{it} is the disturbance for individual i at time t. The sample consists of n individual units.

For a given year, say t_0, let a single cross-section sample on the variables x and y be denoted as

$$y_{1t_0}, y_{2t_0}, \ldots, y_{nt_0}; \; x_{1t_0}, x_{2t_0}, \ldots, x_{nt_0} \quad . \tag{1.2.8}$$

In estimating the parameters β_0 and β_1 in (1.2.7) from this type of sample, it is usually assumed that the individuals in the cross-section sample are homogeneous in behavior in the sense that the ith individual, who associated y_{it_0} with x_{it_0}, would have behaved like the jth individual in associating y_{jt_0} with x_{jt_0} if the value of his x had been x_{jt_0} rather than x_{it_0}. In nonexperimental situations generally encountered in economics it seems unlikely that inter-individual differences observed in a cross-section can be explained by a simple relation with one or a few explanatory variables. One possibility is that the underlying relation between x and y is different for different individuals in the sample. The most plausible reason why different relationships between x and y underlie different individuals' behavior is that some important explanatory variables are missing. For example, if y is household saving and x is household income, it may happen that a young household adjusts its saving to its income along a relation $y_{it_0} = \beta_{0i} + \beta_{1i} x_{it_0} + u_{it_0}$ and an old household adjusts its saving to its income along another relation $y_{jt_0} = \beta_{0j} + \beta_{1j} x_{jt_0} + u_{jt_0}$ such that $\beta_{0i} \neq \beta_{0j}$ and $\beta_{1i} \neq \beta_{1j}$. In this case age is one of the missing variables which accounts for the differences in their behavior. A trivariate equation connecting x,y, and age may have identical parameters for each individual because the inclusion of the age variable brings the households to a comparable level with regard to their saving behavior by eliminating its effect on y. In studying a cross-section sample of households, we must realize that apart from purely economic factors, there are various demographic, sociological, psychological, ane environmental factors which might vary widely among individuals and affect their economic behavior. When we are compelled to work with fewer independent variables than required, it may be desirable to specify that each individual has his own intercept and slope coefficients with regard to included variables. If we have n individuals in our sample we may thus face the problem of interpreting a single relation of the type (1.2.7) as representative in some sense of n underlying equations of the form

$$y_{it_0} = \beta_{0i} + \beta_{1i}x_{it_0} + u_{it_0} \quad (i = 1,2,\ldots,n) \quad . \tag{1.2.9}$$

Once we allow for different intercepts and slopes for each individual, it may be that the remaining differences between their response patterns are attributable just to the u_{it_0}. Thus, the value of specifying a model of the type (1.2.9) might be substantial in econometric work. However, in practice it is difficult to know _a priori_ whether the coefficients vary randomly or systematically across units.

In (1.2.9) suppose that β_{0i}, β_{1i} and u_{it_0} are independently distributed with means β_0, β_1 and zero and with finite positive variances σ_0^2, σ_1^2, and σ_u^2, respectively. It is interesting to note that this set of assumptions covers the familiar fixed coefficient model for a cross-section as a special case. The model (1.2.9) is equivalent to (1.2.7) if β_{0i} and β_{1i} are distributed with zero variances but with finite means β_0 and β_1, respectively. Indeed in the traditional fixed coefficient models, the intercept may be interpreted as precisely a random coefficient: we may consider the quantity $\beta_0 + u_{it_0}$ as a random variable with mean β_0 while the slope coefficient β_1 is fixed. In the more general RCR model of the type (1.2.9), we need to estimate the mean and variance of each coefficient besides the variance of the disturbance term. Thus, the number of parameters to be estimated increases when the coefficients are random. On the other hand, if we treat the coefficient vector in (1.2.9) as fixed but different for different individuals, then the number of coefficient vectors to be estimated grows with the number of units in the sample. Data on each individual unit should be used to estimate its own coefficient vector. This difficulty will not be present if we can assume that the coefficient vector is random but distributed with the same mean and the same variance-covariance matrix.

Hurwicz (1950, pp. 410-8) provided some more examples of RCR models and derived the likelihood function for a RCR model. In deriving the production function, the supply function for factors, and the demand function for a product, Nerlove (1965, pp. 34-35, & Ch. 4) found it appropriate to treat the elasticities of output with respect to inputs and of factor supplies and product demand as random variables differing from firm to firm.

(d) Estimation of RCR Models Using a Single Cross-Section Data

Given a single cross-section sample of the type (1.2.8), suppose we apply least-squares regression of y on x obtaining the slope,

$$\hat{\beta}_1 = \frac{\sum_{i=1}^{n} (x_{it_0} - \bar{x}_{t_0})(y_{it_0} - \bar{y}_{t_0})}{\sum_{i=1}^{n} (x_{it_0} - \bar{x}_{t_0})^2}, \tag{1.2.10}$$

where $\bar{x}_{t_0} = \frac{1}{n}\sum_{i=1}^{n} x_{it_0}$, and $\bar{y}_{t_0} = \frac{1}{n}\sum_{i=1}^{n} y_{it_0}$.

Klein (1953, p. 217) studied the properties of $\hat{\beta}_1$ under the following set of assumptions:

(1) β_{0i} is distributed with mean β_0 and constant variance σ_0^2

(2) β_{1i} is distributed with mean β_1 and constant variance σ_1^2

(3) u_{it_0} is distributed with mean zero and constant variance σ_u^2

(4) β_{0i}, β_{1i}, and u_{it_0} are mutually independent and are independent of themselves for different individuals.

He has concluded that, if $\beta_{0i} + u_{it_0}$ is uncorrelated with x_{it_0} and if β_{1i} is uncorrelated with $x_{it_0}^2$ and $x_{it_0}\bar{x}_{t_0}$, then $\hat{\beta}_1$ is a consistent estimator of β_1. Notice that Klein is treating x_{it_0} as a stochastic variable. If x_{it_0} is a nonstochastic variable, or if x_{it_0} is random and distributed independently of all other random quantities, à la Zellner (1966), it can be easily shown that $\hat{\beta}_1$ is an unbiased estimator of β_1.

Klein also derived nonlinear ML estimating equations for the parameters of the model (1.2.9) assuming that the variable $z_{it_0} = y_{it_0} - \beta_0 - \beta_1 x_{it_0}$ was normally distributed with mean zero and variance $\sigma^2_{z_{it_0}} = \sigma^2_0 + \sigma^2_u + x^2_{it_0}\sigma^2_1$. Notice here that the dependence of the variance of z_{it_0} on x_{it_0} introduced heteroskedasticity in the y_{it_0}'s. Klein could not identify σ^2_0 and σ^2_u separately and thus treated $\sigma^2_0 + \sigma^2_u$ as a single unknown parameter.

Rubin (1950) considered a RCR model of the type

$$y_i = \beta_{0i} + \beta_{1i}x_{1i} + \dots + \beta_{K-1i}x_{K-1i} \qquad (i = 1,2,\dots,n) \qquad (1.2.11)$$

and applied the ML method to estimate its parameters under the restriction that the variances of coefficients are nonnegative. His solution appears to be very complicated. If the observations in (1.2.11) constitute a single cross-section data, the equation (1.2.11) can be obtained by extending (1.2.9) to include K regressors and dropping the disturbance term u_{it_0}. Theil and Mennes (1959) specified a RCR model of the type (1.2.9) omitting the constant term to analyze aggregate time series data on British import and export prices and developed some consistent estimators of its parameters. Subsequently, Hildreth and Houck (1968) extended the Theil and Mennes results and suggested some new consistent estimators. We can represent their model as

$$y_i = \sum_{k=0}^{K-1} x_{ik}(\bar{\beta}_k + \delta_{ik}) \qquad (i = 1,2,\dots,n) \quad , \qquad (1.2.12)$$

where y_i is an observation on a dependent variable, the x_{ik}'s are the observations on K nonstochastic independent variables, the $\bar{\beta}_k$'s are the unknown means of coefficients, the δ_{ik}'s are the additive random elements in coefficients and $x_{i0} = 1$ for all i. The additive disturbance term cannot be distinguished from the randomly varying intercept and so is not written explicitly. It is assumed that the coefficients $\beta_{ik} = \bar{\beta}_k + \delta_{ik}$ fluctuate randomly from one observation to the next such that

$$E\,\delta_{ik} = 0 \quad \text{and} \quad E\,\delta_{ik}\delta_{i'k'} = \begin{cases} \sigma^2_{\delta_k} & \text{if } i = i' \text{ and } k = k' , \\ 0 & \text{otherwise.} \end{cases} \tag{1.2.13}$$

The meaning of the equation (1.2.12) is that if an independent variable increases by one unit, and the other independent variables remain constant, the dependent variable responds with a random change with a finite mean and a positive finite variance. Notice that the specification (1.2.12) is the same as that in (1.2.11). For all the n observations together we can write (1.2.12) as

$$\begin{bmatrix} y_1 \\ y_2 \\ \vdots \\ y_n \end{bmatrix} = \begin{bmatrix} 1 & x_{11} & \cdots & x_{K-1,1} \\ 1 & x_{22} & \cdots & x_{K-1,2} \\ \vdots & \vdots & & \vdots \\ 1 & x_{1n} & \cdots & x_{K-1,n} \end{bmatrix} \begin{bmatrix} \bar{\beta}_0 \\ \bar{\beta}_1 \\ \vdots \\ \bar{\beta}_{K-1} \end{bmatrix} + \begin{bmatrix} \underline{x}'_1 & 0 & \cdots & 0 \\ 0 & \underline{x}'_2 & \cdots & 0 \\ \vdots & \vdots & & \vdots \\ 0 & 0 & \cdots & \underline{x}'_n \end{bmatrix} \begin{bmatrix} \underline{\delta}_1 \\ \underline{\delta}_2 \\ \vdots \\ \underline{\delta}_n \end{bmatrix} \tag{1.2.14}$$

or more compactly as

$$\underline{y} = X\underline{\bar{\beta}} + D_x\underline{\delta} \quad , \tag{1.2.14a}$$

where $\underline{y}$ is a nx1 vector of observations on y_i, X is a nxK matrix of observations with rank K on K independent variables, $\underline{\bar{\beta}}$ is a Kx1 vector of means of coefficients, D_x is a nxKn block diagonal matrix on the r.h.s. of (1.2.14), $\underline{x}'_i$ is the ith row of X and $\underline{\delta}$ is a Knx1 vector of random variables.

An estimator which is the BLUE of $\underline{\bar{\beta}}$ is

$$\hat{\underline{\bar{\beta}}} = (X'H_d^{-1} X)^{-1} X'H_d^{-1} \underline{y} \quad , \tag{1.2.15}$$

where $H_d = D_x(I_n \otimes \Delta_d)D'_x$, $\Delta_d = \text{diag}\,(\sigma^2_{\delta_0}, \sigma^2_{\delta_1}, \ldots, \sigma^2_{\delta_{K-1}})$ and $\otimes$ represents the Kronecker product.

The ith diagonal element of H_d is $\sum_{k=0}^{K-1} \sigma^2_{\delta_k} x^2_{ik}$. This means that we are basically in a heteroskedastic situation as far as the estimation of $\underline{\bar{\beta}}$ is concerned. Since the variances $\sigma^2_{\delta_k}$ are usually unknown, using the squares of the least squares residuals given by $[I - X(X'X)^{-1}X']\underline{y}$, Hildreth and Houck have developed an unrestricted consistent estimator of Δ_d. This is substituted back in (1.2.15) to obtain a consistent estimator of $\bar{\underline{\beta}}$. They have also considered two other estimators of Δ_d because the unrestricted estimator yields negative estimates with positive probability. They suggested a truncated estimator and a quadratic programming estimator to obtain nonnegative estimates for the diagonal elements of Δ_d. Further investigations of sampling properties of these estimators are needed. Note that the above procedure can be adopted either in the analysis of a single cross-section data or in the analysis of a single time series data.

The above procedure can be extended to the general case where the off-diagonal elements of Δ_d are nonzero, i.e., the covariance between every pair of variables $\delta_{i0}, \delta_{i1}, \ldots, \delta_{iK-1}$ is nonzero, cf. Fisk (1967). Such a specification has K unknown means $\bar{\underline{\beta}}$ and (1/2)K(K+1) unknown variances and covariances. Clearly, a sizable sample is needed if all these parameters are to be estimated with any degree of precision. If $\sigma_{\delta_{ik}\delta_{ik'}}$ denotes the covariance between the variables δ_{ik} and $\delta_{ik'}$, $(k \neq k')$, and $\sigma^2_{\delta_{ik}}$ and $\sigma^2_{\delta_{ik'}}$ denote the variances of δ_{ik} and $\delta_{ik'}$ respectively, then nonlinear restrictions of the form $\sigma^2_{\delta_{ik}\delta_{ik'}} \leq \sigma^2_{\delta_{ik}}\sigma^2_{\delta_{ik'}}$ are present and these are difficult to handle. It is possible to impose these restrictions by using methods described by Nelder (1968) for simple cases.

The first author who considered a RCR model was Wald (1947). Assuming that the conditional distribution of the dependent variable y_i given the coefficients $\beta_0, \ldots, \beta_{K-1}$ is normal with mean $\beta_0 x_{0i} + \ldots + \beta_{K-1} x_{K-1,i}$ and a constant variance σ^2, and the coefficients $\beta_0, \ldots, \beta_{K-1}$ are independently and normally distributed with mean zero and a common variance σ_1^2, Wald has constructed a confidence interval for σ_1^2/σ^2.

(e) <u>Aggregation and a RCR Model</u>

In a remarkable note, Zellner (1966) applied a RCR model to the aggregation problem and showed that for such models and for a certain range of specifying assumptions, there would be no aggregation bias in the least squares estimates of coefficients in a macroequation obtained by aggregating a microequation over all microunits. To illustrate his point Zellner made use of the following model. Let the economic relationship for the ith unit be given by

$$\underline{y}_i = X_i\underline{\beta}_i + \underline{u}_i \qquad (i = 1,2,\ldots,n) \quad , \tag{1.2.20}$$

where $\underline{y}_i \equiv (y_{i1}, y_{i2},\ldots, y_{iT})'$ is a T x 1 vector of observations on a dependent random variable, X_i is a T x K matrix of observations with rank K on K nonstochastic independent variables, x_{itk} $(t = 1,2,\ldots,T;\ k = 0,1,\ldots,K-1)$, $\underline{\beta}_i$ is a K x 1 vector of fixed coefficients, $\underline{u}_i \equiv (u_{i1},u_{i2},\ldots,u_{iT})'$ is a T x 1 vector of disturbances with $E\underline{u}_i = 0$ and $E\underline{u}_i\underline{u}_j' = \sigma^2 I$ if $i = j$, 0 if $i \neq j$ and n is the number of units being considered.

Now consider macro variables obtained by simple aggregation; that is,

$$\overline{\underline{y}} = \sum_{i=1}^{n} \underline{y}_i \quad \text{and} \quad \overline{X} = \sum_{i=1}^{n} X_i \quad \text{and} \quad \overline{\underline{u}} = \sum_{i=1}^{n} \underline{u}_i \quad . \tag{1.2.21}$$

Let us postulate the following relationship connecting the macro variables:

$$\overline{\underline{y}} = \overline{X}\underline{\beta} + \underline{\varepsilon} \quad , \tag{1.2.22}$$

where $\overline{\underline{y}}$ and $\overline{X}$ are shown in (1.2.21), $\underline{\beta}$ is a K x 1 vector of fixed macro coefficients and $\underline{\varepsilon}$ is a T x 1 vector of macro disturbances, each with mean zero and variance σ^2.

If we estimate $\underline{\beta}$ in (1.2.22) by least squares using macro data, we get the macro estimator $\hat{\underline{\beta}}$,

$$\hat{\underline{\beta}} = (\overline{X}'\overline{X})^{-1}\overline{X}'\overline{\underline{y}} = (\overline{X}'\overline{X})^{-1}\, \overline{X}' \sum_{i=1}^{n} \underline{y}_i \quad . \tag{1.2.23}$$

Now substitute from (1.2.20) and take expectations as follows:

$$E\hat{\underline{\beta}} = E(\overline{X}'\overline{X})^{-1}\overline{X} \sum_{i=1}^{n} (X_i\underline{\beta}_i + \underline{u}_i) = \sum_{i=1}^{n} (\overline{X}'\overline{X})^{-1}\, \overline{X}'X_i\underline{\beta}_i \quad . \tag{1.2.24}$$

Notice here that $\sum_{i=1}^{n} (\overline{X}'\overline{X})^{-1}\overline{X}'X_i = I$. It will be recognized that $(\overline{X}'\overline{X})^{-1}\overline{X}'X_i$ is the matrix of coefficients in the "auxiliary regressions" of the X_i upon $\overline{X}$. It is clear from (1.2.24) that in general the expectation of a single element of $\underline{\beta}$ will depend on corresponding and noncorresponding micro coefficients. To evaluate the aggregation bias in $\underline{\beta}$ we have to relate it to the "true value" of macro parameter vector which can be defined by $\frac{1}{n}\sum_{i=1}^{n}\underline{\beta}_i$. With this definition we can rewrite (1.2.24) as

$$E\hat{\underline{\beta}} = \sum_{i=1}^{n} [(\overline{X}'\overline{X})^{-1}\overline{X}'X_i - n^{-1}I]\,\underline{\beta}_i + n^{-1}\sum_{i=1}^{n}\underline{\beta}_i \quad . \tag{1.2.25}$$

Since $n^{-1}I$ is the mean matrix of $(\overline{X}'\overline{X})^{-1}\overline{X}'X_i$, the first term on the r.h.s. of (1.2.25) will be recognized as n times the variance-covariance matrix of the weights $(\overline{X}'\overline{X})^{-1}\overline{X}'X_i$ with the micro parameters $\underline{\beta}_i$. When this variance-covariance matrix is null, we satisfy Theil's (1954, p. 16) perfect aggregation condition.

At this point let us turn to Zellner's approach to the aggregation problem. Following Zellner, let

$$\underline{\beta}_i = \overline{\underline{\beta}} + \underline{\delta}_i \quad , \tag{1.2.26}$$

where $\underline{\delta}_i$ is a random vector with $E\underline{\delta}_i = 0$.

Using (1.2.26) we can rewrite (1.2.20) as

$$\underline{y}_i = X_i\overline{\underline{\beta}} + X_i\underline{\delta}_i + \underline{u}_i \qquad (i = 1,2,\ldots,n) \quad . \tag{1.2.27}$$

By summing (1.2.27) over i we get

$$\overline{\underline{y}} = \overline{X}\overline{\underline{\beta}} + \sum_{i=1}^{n} X_i\underline{\delta}_i + \overline{\underline{u}} \quad . \tag{1.2.28}$$

Substituting (1.2.28) in (1.2.23) we have

$$\hat{\underline{\beta}} = (\overline{X}'\overline{X})^{-1}\overline{X}'\left(\overline{X}\overline{\underline{\beta}} + \sum_{i=1}^{n} X_i\underline{\delta}_i + \overline{\underline{u}}\right) \quad . \tag{1.2.29}$$

Taking expectation on both sides of (1.2.29) we get

$$E\hat{\underline{\beta}} = \overline{\underline{\beta}} \quad , \tag{1.2.30}$$

because $E(\overline{X}'\overline{X})^{-1}\overline{X}'\sum_{i=1}^{n} X_i\underline{\delta}_i = 0$, and $E(\overline{X}'\overline{X})^{-1}\overline{X}'\overline{\underline{u}} = 0$.

Thus, there is no aggregation bias in $\hat{\underline{\beta}}$. Through the assumption (1.2.26) and the assumption that the explanatory variables are nonstochastic, Theil's (1954) special case of no correlation between micro coefficients $\underline{\beta}_i$ and the weights $(\bar{X}'\bar{X})^{-1}\bar{X}'X_i$ shown in (1.2.25) has essentially been built into the model.

If the elements of X_i are stochastic and if they are distributed independently of the $\underline{\delta}_i$ and $\underline{u}_i$, $\hat{\underline{\beta}}$ is still an unbiased estimator of $\bar{\underline{\beta}}$. If $\bar{X}$ contains lagged values of $\bar{\underline{y}}$, (1.2.30) can be reformulated to read plim $\hat{\underline{\beta}} = \bar{\underline{\beta}}$.

The sampling error of the estimator $\underline{\beta}$ relative to $\bar{\underline{\beta}}$ is given by

$$\hat{\underline{\beta}} - \bar{\underline{\beta}} = (\bar{X}'\bar{X})^{-1}\bar{X}' \sum_{i=1}^{n} X_i\underline{\delta}_i + (\bar{X}'\bar{X})^{-1}\bar{X}'\bar{\underline{u}} \quad , \tag{1.2.31}$$

which reflects two sources of randomness, namely that arising from the $\underline{\delta}_i$ and that arising from the $\bar{\underline{u}}$.

It has been noted by Theil (1971, Chapter 11) that for sufficiently large n we can replace the equation

$$\frac{1}{n}\sum_{i=1}^{n} \underline{y}_i = \frac{1}{n}\sum_{i=1}^{n} X_i\bar{\underline{\beta}} + \frac{1}{n}\sum_{i=1}^{n} X_i\underline{\delta}_i + \frac{1}{n}\sum_{i=1}^{n} \underline{u}_i \tag{1.2.28a}$$

by

$$\frac{1}{n}\sum_{i=1}^{n} \underline{y}_i \approx \frac{1}{n}\sum_{i=1}^{n} X_i\bar{\underline{\beta}} + \frac{1}{n}\sum_{i=1}^{n} \underline{u}_i \quad ,$$

if the variance-covariance matrix of $\frac{1}{n}\sum_{i=1}^{n} X_i\underline{\delta}_i$ converges to a null matrix as n increases indefinitely and T is fixed. Similarly, it may also happen that the variance-covariance matrix of the disturbance term $\frac{1}{n}\sum_{i=1}^{n} \underline{u}_i$ tends to a null matrix as $n \to \infty$ and T is fixed. In this case we can also remove the disturbance term from (1.2.28a) when n is sufficiently large. To avoid this difficulty we may write $\underline{u}_i = \underline{\tau} + \underline{\nu}_i$ where $\underline{\tau} \equiv (\tau_1, \tau_2, \ldots, \tau_T)'$, $\underline{\nu}_i \equiv (\nu_{i1}, \nu_{i2}, \ldots, \nu_{iT})'$, τ_t and ν_{it} are as explained in (1.2.6). If the vectors $\frac{1}{n}\sum_{i=1}^{n} \underline{\nu}_i$ and $\frac{1}{n}\sum_{i=1}^{n} X_i\underline{\delta}_i$ converge in probability to a null vector as $n \to \infty$ and T is fixed we can write (1.2.28a) for sufficiently large n as

$$\frac{1}{n}\sum_{i=1}^{n} \underline{y}_i \approx \frac{1}{n}\sum_{i=1}^{n} X_i\underline{\bar{\beta}} + \underline{\tau} \quad ,$$

which is a traditional fixed coefficient model.

Notice that the RCR model (1.2.27) adopted by Zellner is the same as that in Rao (1965a, p. 192) but is different from the model analyzed by Klein, Theil and Mennes and others referred to in subsection 1.2(d). The coefficient vector in (1.2.27) varies randomly across units. Once a cross-section unit is selected, a drawing on its coefficient vector is kept the same for all the time series observations on that unit. Each cross-section unit is assumed to have a stable behavior over time. If the population consists of a large number of units whose behavior is described by a relation, then the randomness of coefficients in that relation may be attributed to the random selection of units. If we observe the sample units over a short period of time, we may reasonably assume that the units have stable behavior over time. In Chapter IV we cite some economic examples for which Zellner's specification is appropriate. In (1.2.12) different drawings on coefficients are associated with different observations on the independent variables. In this sense the specification (1.2.12) is more general than (1.2.27). The former model is also of considerable interest to economists. If we want to use the model (1.2.12) in the analysis of panel data, we have to vary the coefficients both over time and across units. That is, we have to modify (1.2.27) as

$$y_{it} = \beta_{0it} + \beta_{1it}x_{1it} + \dots + \beta_{K-1,it}x_{K-1,it} \tag{1.2.12a}$$

$$(i = 1,2,\dots,n;\ t = 1,2,\dots,T) \quad .$$

Fisk (1967) has studied a RCR model of the type (1.2.12a) under the assumption that the coefficient vectors $\underline{\beta}_{it} = (\beta_{0it},\beta_{1it},\dots,\beta_{K-1,it})'$ are independently and identically distributed. If the coefficient vectors are serially correlated and the units are randomly drawn, then this assumption is not tenable. Rosenberg (1970) also studied a RCR model of the type (1.2.12a) but he followed a much more realistic approach. He has adopted a convergent-parameter structure, in which individual coefficient vectors $\underline{\beta}_{it}$ evolve stochastically over time, exhibiting a tendency to converge toward the population mean vector and undergoing random shifts as well. The population mean vector also evolves stochastically over time in this scheme. We need lots of prior information to estimate the model (1.2.12a) under these assumptions.

Although the term $(\beta_0 + \mu_i + \tau_t + \upsilon_{it})$ in (1.2.6) varies randomly in both the time and cross-section dimensions when the μ_i's and τ_t's are random variables, we cannot say that (1.2.6) is a particular case of (1.2.12a) because the term τ_t introduces correlation between the terms $(\beta_0+\mu_i+\tau_t+\upsilon_{it})$ and $(\beta_0+\mu_j+\tau_t+\upsilon_{jt})$ $(i \neq j)$ in (1.2.6) and no such correlation between β_{0it} and β_{0jt} for $i \neq j$ in (1.2.12a) is apparent. If the specific year effects are in general present, we are not justified in assuming that β_{0it} in (1.2.12a) are independent of themselves for different individuals.

Suppose that it is reasonable to assume that the K-dimensional coefficient vectors $\underline{\beta}_{it}$ $(i = 1,2,\ldots,n;\ t = 1,2,\ldots,T)$ in (1.2.12a) are independently and identically distributed with the common mean vector $\bar{\underline{\beta}} \equiv (\bar{\beta}_0, \bar{\beta}_1,\ldots,\bar{\beta}_{K-1})'$ and the common variance-covariance matrix Δ. Suppose that the K-dimensional coefficient vectors $\underline{\beta}_i = (\beta_{0i},\ldots,\beta_{K-1,i})'(i = 1,2,\ldots,n)$ in (1.2.27) are independently and identically distributed with the common mean vector $\bar{\underline{\beta}} \equiv (\bar{\beta}_0,\bar{\beta}_1,\ldots,\bar{\beta}_{K-1})'$ and the common variance-covariance matrix Δ. Let $\underline{u}_i$ be independent of $\underline{\beta}_j$. Under these assumptions, it is instructive to compare the mean and variance-covariance matrix of $\underline{y}_i$ in (1.2.27) with those of $\underline{y}_i \equiv (y_{i1},y_{i2},\ldots,y_{iT})'$ in (1.2.12a). The mean of $\underline{y}_i$ in either case is the same. The variance-covariance matrix of $\underline{y}_i$ in (1.2.27) is $X_i\Delta X_i' + \sigma^2 I$ and that of $\underline{y}_i$ in (1.2.12a) is diag $[\underline{x}_{i1}'\Delta\underline{x}_{i1}, \underline{x}_{i2}'\Delta\underline{x}_{i2},\ldots,\underline{x}_{iT}'\Delta\underline{x}_{iT}]$ where $\underline{x}_{it}'$ is the tth row of X_i. This shows that the variance of an observation on the dependent variable in (1.2.27) exceeds that of a corresponding observation on the dependent variable in (1.2.12a) by a quantity σ^2 but the time series observations on the dependent variable in (1.2.27) for an individual unit are correlated unlike those on the dependent variable in (1.2.12a). The distribution which generated the observations on the dependent variable in (1.2.12a) is not more general than that which generated the observations on the dependent variable in (1.2.27) even though the RCR model (1.2.12a) is specified in a general way. The estimation and testing procedures within the framework of the model (1.2.12) or (1.2.12a) are highly complicated. Zellner's approach provides a way of simplifying the statistical inference in RCR models. We can make use of standard multivariate techniques to analyze (1.2.27). This point will be clear from our discussion in Chapter IV. The specification (1.2.27) is well suited to the analysis of time series of cross-sections. If all the independent variables in (1.2.4) are fixed, then the model (1.2.4) is a particular case of (1.2.27).

Zellner's assumption that coefficient vectors of different individuals are the independent drawings from the same multivariate distribution stands between the limiting assumptions that coefficient vectors are fixed and the same and that coefficient vectors are fixed and different. Given that micro units are possibly different in their behavior, the former assumption is often found to be restrictive while the latter assumption involves the use of very many parameters and is thus not always satisfactory in the analysis of panel data pertaining to many individuals.

(f) Estimation of RCR Models Using Replicated Data[1]

Rao (1965b) proves a modified Gauss-Markov theorem on least squares when the parameters are stochastic. A new vista within the sampling theory framework has been opened up in his results on the estimation and testing of random coefficient models. He considers the following model:

$$\underline{y}_i = X_i \underline{\beta}_i + \underline{u}_i \qquad (i = 1,2,\ldots,n) \tag{1.2.32}$$

where $\underline{y}_i$ is a Tx1 vector of observations on the dependent variable for the ith

---

[1] Tiao (1966) applied a Bayesian approach to the problem of analyzing the random effects in a regression model with variance components appearing in the error term. Tiao incorporated a random intercept keeping the slopes fixed.

individual, X_i is a TxK matrix of observations with rank K on K nonstochastic independent variables. $\underline{\beta}_i$ is a Kx1 vector of coefficient varying randomly across units and $\underline{u}_i$ is a Tx1 vector of disturbances. Notice that the model (1.2.32) is the same as (1.2.27).

Rao confined his attention to the experimental situation in which the "design matrix" X was kept constant for all individuals. That is, each individual unit is fed with the same matrix of observations on independent variables. Therefore, let

$$X_1 = X_2 = \ldots = X_n = \bar{X} \quad . \qquad (1.2.33)$$

He further averaged the relation (1.2.32) over all individuals under the assumption that $\underline{y}_i$ and $\underline{y}_j$ for $i \neq j$ are independent. He then rewrote (1.2.32) as

$$\bar{\underline{y}} = \bar{X}\underline{\beta} + \bar{\underline{u}} \quad , \qquad (1.2.34)$$

where $\bar{\underline{y}}$, $\underline{\beta}$, $\bar{\underline{u}}$ are simple averages of y_i's, $\underline{\beta}_i$'s, and $\underline{u}_i$'s, respectively.[2]

Estimation of parameters in (1.2.34) was considered under the following assumptions:

(1) $E\bar{\underline{u}} = 0,\ E\bar{\underline{u}}\,\bar{\underline{u}}' = n^{-1}\Sigma$; (1.2.35a)

(2) $E\underline{\beta} = \bar{\underline{\beta}},\ E(\underline{\beta}-\bar{\underline{\beta}})(\underline{\beta}-\bar{\underline{\beta}})' = n^{-1}\Delta$; (1.2.35b)

(3) $\text{cov}(\underline{\beta}, \bar{\underline{u}}) = 0$. (1.2.35c)

Rao proved three lemmas on the estimation of linear functions of the elements of random vector $\underline{\beta}$ and of its mean vector $\bar{\underline{\beta}}$ under the above assumptions. We state his lemmas below without proof since they are referred to in our subsequent work.

Lemma 1.2.1: Let $\underline{\ell}'\underline{\beta}$ be a linear function of $\underline{\beta}$ and $a + \underline{L}'\bar{\underline{y}}$ be a linear predictor of $\underline{\ell}'\underline{\beta}$. The minimum variance linear unbiased predictor of $\underline{\ell}'\underline{\beta}$--i.e., the function $a + \underline{L}'\bar{\underline{y}}$ which minimizes $E(\underline{\ell}'\underline{\beta} - a - \underline{L}'\bar{\underline{y}})^2$ subject to

[2] A simple average of the relation is appropriate here because of the condition (1.2.33). The question of taking a weighted average of the relation is relevant when the matrix X_i is different for different individuals. We will discuss this point in detail in Chapter IV.

$E(\underline{\ell}'\underline{\beta} - a - \underline{L}'\underline{y}) = 0$ -- is determined by

$$a = 0,\ \underline{L} = \Sigma^{-1}\bar{X}(\bar{X}'\Sigma^{-1}\bar{X})^{-1}\underline{\ell} \quad .$$

Its prediction variance is

$$E[\underline{\ell}'\underline{\beta}-\underline{\ell}'(\bar{X}'\Sigma^{-1}\bar{X})^{-1}\bar{X}'\Sigma^{-1}\bar{\underline{y}}]^2 = n^{-1}\underline{\ell}'(\bar{X}'\Sigma^{-1}\bar{X})^{-1}\underline{\ell} \quad .$$

Lemma 1.2.2: Let $\underline{\ell}'\bar{\underline{\beta}}$ be a linear function of $\bar{\underline{\beta}}$ and $a + \underline{L}'\bar{\underline{y}}$ be a linear estimator of $\underline{\ell}'\bar{\underline{\beta}}$. The minimum variance linear unbiased estimator of $\underline{\ell}'\bar{\underline{\beta}}$ -- i.e., the function $a + \underline{L}'\bar{\underline{y}}$ which minimizes $E(\underline{\ell}'\bar{\underline{\beta}} - a - \underline{L}'\bar{\underline{y}})^2$ subject to $E(\underline{\ell}'\bar{\underline{\beta}} - a - \underline{L}'\bar{\underline{y}}) = 0$ -- is determined by

$$a = 0,\ \underline{L} = \Sigma^{-1}\bar{X}(\bar{X}'\Sigma^{-1}\bar{X})^{-1}\underline{\ell} \quad .$$

Its variance is

$$E[\underline{\ell}'\bar{\underline{\beta}} - \underline{\ell}'(\bar{X}'\Sigma^{-1}\bar{X})^{-1}\bar{X}'\Sigma^{-1}\bar{\underline{y}}]^2 = n^{-1}\underline{\ell}'\Delta\underline{\ell} + n^{-1}\underline{\ell}'(\bar{X}'\Sigma^{-1}\bar{X})^{-1}\underline{\ell} \ .$$

From here on Rao considered the special case where $\Sigma = \sigma^2 I$. An unbiased estimator of σ^2 is

$$\hat{\sigma}^2 = \frac{tr[I-\bar{X}(\bar{X}'\bar{X})^{-1}\bar{X}'](S + n\,\bar{\underline{y}}\bar{\underline{y}}')}{n(T-K)} \quad , \qquad (1.2.36)$$

where $S = \sum_{i=1}^{n} \underline{y}_i\underline{y}_i' - n\bar{\underline{y}}\bar{\underline{y}}'$, the matrix of corrected sum of squares and products.

An unbiased estimator of Δ is

$$\hat{\Delta} = \frac{(\bar{X}'\bar{X})^{-1}\bar{X}'S\bar{X}(\bar{X}'\bar{X})^{-1}}{n-1} - \hat{\sigma}^2(\bar{X}'\bar{X})^{-1} \quad . \qquad (1.2.37)$$

An unbiased estimator of $\bar{\underline{\beta}}$ is

$$\hat{\bar{\underline{\beta}}} = (\bar{X}'\bar{X})^{-1}\bar{X}'\bar{\underline{y}} \quad , \qquad (1.2.38)$$

which is simply the mean of least squares estimators of $\underline{\beta}_i$ from all the n equations in (1.2.32).

Writing $S + n\bar{\underline{y}}\bar{\underline{y}}' = \sum_{i=1}^{n} \underline{y}_i\underline{y}_i'$, substituting $\bar{X}\underline{\beta}_i + \underline{u}_i$ for $\underline{y}_i$ and making use of the result $[I-\bar{X}(\bar{X}'\bar{X})^{-1}\bar{X}']\bar{X} = 0$, we have

$$\hat{\sigma}^2 = \sum_{i=1}^{n} \frac{\hat{\underline{u}}_i'\hat{\underline{u}}_i}{n(T-K)} \quad , \tag{1.2.39}$$

where $\hat{\underline{u}}_i$ is the vector of least squares residuals from the ith regression in (1.2.32). Thus, the estimation of σ^2 using the estimator in (1.2.36) is equivalent to adding the sum of squares of least squares residuals in each equation across units, and dividing it by the overall degrees of freedom (≡ d.f.).

We can also write

$$\hat{\Delta} = \frac{1}{n-1}[(\bar{X}'\bar{X})^{-1}\bar{X}'\left\{\sum_{i=1}^{n} \underline{y}_i\underline{y}_i' - n\bar{\underline{y}}\bar{\underline{y}}'\right\}\bar{X}(\bar{X}'\bar{X})^{-1}] - \hat{\sigma}^2(\bar{X}'\bar{X})^{-1}$$

$$= \frac{1}{n-1}\left[\sum_{i=1}^{n} \underline{b}_i\underline{b}_i' - \frac{1}{n}\sum_{i=1}^{n}\underline{b}_i \sum_{i=1}^{n}\underline{b}_i'\right] - \hat{\sigma}^2(\bar{X}'\bar{X})^{-1} \quad , \tag{1.2.40}$$

where $\underline{b}_i = (\bar{X}'\bar{X})^{-1}\bar{X}'\underline{y}_i$.

This shows that $\underline{\Delta}$ can be estimated taking the sample variance--covariance matrix of least squares quantities $\underline{b}_i$ and subtracting from it the expression $\hat{\sigma}^2(\bar{X}'\bar{X})^{-1}$.

The problems of testing and construction of confidence intervals for $\underline{\ell}'\underline{\beta}$ and $\underline{\ell}'\bar{\underline{\beta}}$ are examined by Rao under the following additional assumptions. When appropriate we can assume that $\underline{\beta}$ is distributed as K-dimensional normal with mean $\bar{\underline{\beta}}$ and variance-covariance matrix $n^{-1}\Delta$, and $\bar{\underline{u}}$ is distributed as T-dimensional normal with mean $\underline{0}$ and variance-covariance matrix $n^{-1}\sigma^2 I$,[3] i.e.,

$$\underline{\beta} \sim N_K(\bar{\underline{\beta}}, n^{-1}\Delta) \quad \text{and} \quad \bar{\underline{u}} \sim N_T(\underline{0}, n^{-1}\sigma^2 I) \quad . \tag{1.2.41}$$

[3] In some models the coefficients do not take values between $-\infty$ and ∞ but in some specific ranges. For example, the marginal propensity to consume takes values between 0 and 1. In these cases it is not reasonable to assume that the coefficient vector is distributed as normal across units unless the assumed normal distribution is tight in such a way that the relevant region contains most of its probability mass.

From the assumptions (1.2.35) and (1.2.41) it follows that $E\frac{S}{n-1} = \bar{X}\Delta\bar{X}' + \sigma^2 I$ and that S is Wishart distributed with n-1 d.f. provided $n-1 \geq T$. Further, S, $\bar{\underline{u}}$, $\underline{\beta}$ are all stochastically independent. Using the likelihood ratio criterion he develops a test to examine on the basis of S whether the dispersion matrix of $\underline{y}_i$ is of the form $\bar{X}\Delta\bar{X}' + \sigma^2 I$ and another test to examine on the basis of $\bar{\underline{y}}$ and S whether the $E\underline{y}_i = \bar{X}\bar{\underline{\beta}}$.

Rao's theory rests heavily on the possibility of obtaining panel data in a manner that the observational vectors on the regressand can be treated as following the same multivariate distribution with a finite mean vector and a finite variance-covariance matrix. In the case of the above model the n observational vectors on $\underline{y}_i$ are treated as a random sample of size n drawn from a population with mean $\bar{X}\underline{\beta}$ and variance-covariance matrix $\bar{X}\Delta\bar{X}' + \sigma^2 I$.

Before closing this section we should make another reference to Fisk's (1967) article. He considered the following model with variable coefficients

$$y_{it} = \underline{x}'_{it}\underline{\beta}_{it} \qquad (i = 1,2,\ldots,n; \quad t = 1,2,\ldots,T) \quad , \qquad (1.2.42)$$

where y_{it} is an observation on the dependent variable, $\underline{x}'_{it}$ is a 1 x K vector of observations on K nonstochastic independent variables and $\underline{\beta}_{it}$ is a K x 1 vector of coefficients varying randomly both over time and across units. Thus, Fisk's specification is the same as that in (1.2.12a). Fisk sets $\underline{x}'_{it} = \underline{x}'_t$ for all i because in experimental situations repeated observations on the dependent variable for a year t can be obtained fixing the vector of observations on K independent variables at a preselected level. Experimental situations have been considered throughout the bulk of the paper. If n or T is equal to 1 and we assume, ab initio, that the variance-covariance matrix of the coefficient vector is diagonal, then we can follow the approach developed by Hildreth and Houck (1968) to estimate the parameters of the model (1.2.42) without any restriction on $\underline{x}'_{it}$.

1.3 Conclusions

As indicated by Kuh (1959,1963), Mundlak (1961,1963), Hoch (1962), Nerlove (1965), and others, an approach to analyze the combined cross-section and time series data is to specify a regression equation with additional effects associated with both time direction and cross-sectional units. If these effects are random the intercept of the equation will be random. An interesting extension of these specifications is to consider a regression equation with all coefficients as random. This approach is particularly useful in the analyses of panel data if the interindividual parameter variation takes a more general form than mere shifts in the regression intercept. As observed by Klein, specification of a RCR model is appealing in the analysis of cross-section data since it permits corresponding coefficients to be different for different individuals. While specifying a RCR model we can either follow Zellner (1966) and vary the coefficients randomly across units keeping a drawing on a coefficient vector the same for all the time series observations on the variables or follow Theil and Mennes (1959) and vary the coefficients randomly from one observation to the next. Our contribution in this thesis is relevant only to Zellner's specification.

It is unrealistic to make an assumption of the type (1.2.33) in econometric work because in numerous nonexperimental situations encountered by economists the variables treated as dependent and independent variables are actually observed. We can get T observations on each of the variables for a single unit, but it is hard to get repeated observations on a dependent variable for additional individual units, which have the same mean ($\bar{X}\underline{\beta}$) and the same variance-covariance matrix ($\bar{X}\Delta\bar{X}' + \sigma^2 I$) as the dependent variable for the first individual unit, because the matrix of observations on independent variables will not be identical across individual units. Therefore, Rao's theory will be modified in the subsequent chapter to allow for differing matrices of observations on independent variables for different individual units.

CHAPTER II

EFFICIENT METHODS OF ESTIMATING A REGRESSION EQUATION WITH EQUICORRELATED DISTURBANCES

2.1 Introduction

In this chapter we continue the discussion which we initiated in subsection 1.2(a). We consider a linear regression model whose disturbances have a constant variance and pairwise equal correlation. In analyzing this model we exploit the fact that if the off-diagonal elements of the variance-covariance matrix of the disturbances in a regression are all equal, then there is an orthogonal matrix that transforms the problem into one of heteroskedasticity. One of the transformed observations will have a variance different from those of the remaining transformed observations, which will all have the same variance. When these variances are unknown, the choice is open to drop the observation with the higher variance, and that if the variance is enough higher it should be dropped.

This model with equicorrelated disturbances is related to the model (1.2.4) with random intercept and fixed slopes and is useful in analyzing the combined time series and cross-section data, where the various observations for a single unit are equicorrelated because of a unit effect shown in (1.2.2). Then the transformation can be applied separately to the data for each unit. After pooling, then, one has the following problem of heteroskedasticity: n observations with high and equal variances, and (n-1)T observations with low and equal variances. When these variances are unknown, the choice is again open to drop the n observations with high variances.

In Section 2.2 we state certain matrix results which will be used later on in this chapter. In Section 2.3 we discuss an optimal way of using the available data on a single individual unit to estimate the parameters of a regression equation with equicorrelated disturbances. In Section 2.4 we relate this model to another model with random intercept and fixed slopes. We use the latter model in the analysis of combined time series and cross-section data and develop different estimators of its coefficient vector. We study both the small sample and the large sample properties of these estimators and indicate the conditions under which these estimators are optimal according to the minimum variance criterion. In Section 2.5 we point out that when the correlation between every pair of elements of the disturbance vector pertaining to every randomly selected unit is equal to unity, an optimal procedure of estimating the coefficient vector from the combined time series and cross-section data is the OLS procedure under certain exact linear restrictions on coefficients. In Section 2.6 we turn to the problem of estimating the parameters when the intercept is random across units, slopes are fixed and the same for all units, and the error variances vary across units.

2.2 Some Useful Lemmas

Now we will state some well known matrix results on which we rely in our subsequent discussion in this chapter.

Lemma 2.2.1: Let $\Omega = (1-\rho)I_T + \rho\underline{\iota}_T\underline{\iota}_T'$ where $\underline{\iota}_T$ is a T x 1 vector of unit elements and $-(T-1)^{-1} < \rho < 1$. Then

(i) $|\Omega| = (1-\rho)^{T-1}(1-\rho+T\rho)$

(ii) $\Omega^{-1} = (1-\rho)^{-1}I_T - \rho(1-\rho)^{-1}(1-\rho+T\rho)^{-1}\,\underline{\iota}_T\underline{\iota}_T'$

(iii) the eigen values of Ω are $(1-\rho+T\rho)$ with the eigen vector $\underline{\iota}_T'$ and $(1-\rho)$ of multiplicity $(T-1)$ with $(T-1)$ eigen vectors orthogonal to $\underline{\iota}_T'$.

The proof is straightforward and is therefore omitted, cf. Rao (1965a, pp. 53-54, Problem 1.1).

Lemma 2.2.2: Let A be a nonsingular matrix of order T, and $\underline{\ell}_1$ and $\underline{\ell}_2$ be two T x 1 vectors. Then

$$(A + \underline{\ell}_1\underline{\ell}_2')^{-1} = A^{-1} - A^{-1}\underline{\ell}_1\underline{\ell}_2'\,A^{-1}(1+\underline{\ell}_2'\,A^{-1}\underline{\ell}_1)^{-1} \quad .$$

We can easily verify the above result by post multiplying both sides by $(A + \underline{\ell}_1\underline{\ell}_2')$, cf. Rao (1965a, p. 29, Problem 2.8).[5]

Lemma 2.2.3: Let A and B be real T x T symmetric matrices of which B is positive definite. Then there exists a nonsingular matrix P such that $P'AP = \Lambda$, a diagonal matrix and $P'BP = I$. Further, the diagonal elements of Λ are the roots of $|A - \lambda B| = 0$.

The proof is given in Rao (1965a, p. 37).

Lemma 2.2.4: Let A and B be two T x T real, symmetric matrices. Then $|A - \lambda B| = 0$ has all its roots $\lambda \geq 1$ if B is positive definite and (A - B) is nonnegative definite, and has all its roots $0 < \lambda < 1$ if A and B are positive definite matrices and (A - B) is negative definite.

The proof is given in Fisher (1966, p. 187).

Lemma 2.2.5: Let A be a symmetric matrix and B be a positive definite matrix. Denote by $\lambda_1 \geq .... \geq \lambda_m$, the roots of $|A - \lambda B| = 0$. Then

$$\sup_{\underline{\ell}} \frac{\underline{\ell}'A\underline{\ell}}{\underline{\ell}'B\underline{\ell}} = \lambda_1 \quad \text{and} \quad \inf_{\underline{\ell}} \frac{\underline{\ell}'A\underline{\ell}}{\underline{\ell}'B\underline{\ell}} = \lambda_m \quad .$$

The proof of this lemma is given in Trac y and Dwyer (1969, p. 1585).

2.3 A Regression Model with Equicorrelated Disturbances

Let the economic relationship for an ith unit, say firm or household, be given by

$$\underline{y}_i = \beta_0\underline{\iota}_T + Z_i\underline{\delta} + \underline{u}_i \quad , \qquad (2.3.1)$$

---

[5]Rao's statement contains a slight mistake. Our Lemma 2.2.2 provides the correct statement of the result.

where $\underline{y}_i \equiv (y_{i1}, y_{i2}, \ldots, y_{iT})'$ is a T x 1 vector of observations on the dependent variable, β_0 is a constant term, $\underline{\iota}_T \equiv (1,1,\ldots,1)'$ is a T x 1 vector of unit elements, Z_i is a T x (K-1) matrix of observations on K - 1 independent variables, $z_{itk} (t = 1,2,\ldots,T;\ k = 1,2,\ldots,K-1)$; $X_i \equiv [\underline{\iota}_T, Z_i]$ is a T x K matrix, $\underline{\delta} \equiv (\delta_1, \delta_2, \ldots, \delta_{K-1})'$ is a (K-1) x 1 vector of slope coefficients and $\underline{u}_i \equiv (u_{i1}, u_{i2}, \ldots, u_{iT})'$ is a T x 1 vector of disturbances. The subscript t refers to the value of a variable for the year t. With T time series observations on each variable for a single unit we try to estimate the parameters of the model (2.3.1) under the following assumption. Let $\underline{\beta} \equiv [\beta_0, \underline{\delta}']'$.

Assumption 2.3.1:

(1) The sample size T is greater than K.

(2) X_i is fixed in repeated samples on $\underline{y}_i$ and the rank of X_i is K.

(3) $E\underline{u}_i = 0$ and $E\underline{u}_i\underline{u}_i' = \sigma^2\Omega$ where $\Omega = (1-\rho)I_T + \rho\underline{\iota}_T\underline{\iota}_T'$ and $-(T-1)^{-1} < \rho < 1$.

The result (i) in Lemma 2.2.1 indicates that Ω is singular when $\rho = 1$ or $\rho = -(T-1)^{-1}$. The above restriction on the range of ρ is imposed to avoid the cases where Ω is singular.

Application of Aitken's generalized least squares to (2.3.1) will yield, as is well known, a BLUE which is

$$\underline{b}_i(\rho) = (X_i'\Omega^{-1}X_i)^{-1}X_i'\Omega^{-1}\underline{y}_i \quad . \qquad (2.3.2)$$

According to Rao (1967, 1968) a necessary and sufficient (n & s) condition that the OLS estimator

$$\underline{b}_i = (X_i'X_i)^{-1}X_i'\underline{y}_i \quad , \qquad (2.3.3)$$

is the same as the Aitken estimator, $\underline{b}_i(\rho)$ with probability unity is that

$$\sigma^2 X_i'\Omega L_i = 0 \qquad (2.3.4a)$$

or equivalently, $\sigma^2\Omega$ be expressible in the form

$$X_i \Gamma X_i' + L_i \theta L_i' + \omega^2 I_T \tag{2.3.4b}$$

where Γ and θ are arbitrary symmetric matrices, ω^2 is an arbitrary scalar and L_i is a T x (T-K) matrix of rank T-K such that $X_i'L_i = 0$. Some further results and bibliography on this subject are given by Zyskind (1967, 1969) and Watson (1968). Recognizing that the condition $X_i'L_i = 0$ imples $\underline{\iota}_T'L_i = 0$ when $\underline{\iota}_T$ is a column of X_i, we can easily show that Ω satisfies the n & s condition (2.3.4a). Since $\underline{\iota}_T$ is the first column of X_i, we can write $\underline{\iota}_T = X_i\underline{i}_1$ where $\underline{i}_1$ is the first column of an identity matrix of order K. Consequently, we can write $\Omega = \rho X_i \underline{i}_1 \underline{i}_1' X_i' + (1-\rho) I_T$. With $\sigma^2 = 1$, $\Gamma = \rho \underline{i}_1 \underline{i}_1'$, $\theta = 0$ and $\omega^2 = 1 - \rho$, this form for Ω is the one given in (2.3.4b). Thus, when Assumption 2.3.1 is true and a constant term is included in a regression, the n & s conditions for an OLS estimator to be the same as an Aitken estimator in any given sample are satisfied. In this case the knowledge of ρ is not required to obtain the BLUE of $\underline{\beta}$.

Suppose that the equation (2.3.1) does not contain a constant term, i.e., $\beta_0 = 0$, and Assumption 2.3.1 is true. Now the OLS estimator is not an efficient estimator. The BLUE of $\underline{\delta}$ is the Aitken estimator, $\underline{d}_i(\rho) = (Z_i'\Omega^{-1}Z_i)^{-1}Z_i'\Omega^{-1}\underline{y}_i$, which involves ρ. If ρ is unknown, as it usually is, we cannot use the Aitken estimator in practice. In quest of an operational procedure we consider an orthogonal matrix which diagonalizes Ω. From the result (iii) in Lemma 2.2.1 it follows that such an orthogonal matrix will have $\underline{\iota}_T'$ as its first row. Let $0_T = [\underline{\iota}_T/\sqrt{T}, C_1']'$ be an orthogonal matrix which diagonalizes Ω. Premultiplying (2.3.1) by 0_T we have

$$\begin{bmatrix} \bar{y}_i \\ \\ \tilde{\underline{y}}_i \end{bmatrix} = \begin{bmatrix} \bar{\underline{z}}_i' \\ \\ \tilde{Z}_i \end{bmatrix} \underline{\delta} + \begin{bmatrix} \bar{u}_i \\ \\ \tilde{\underline{u}}_i \end{bmatrix} \tag{2.3.5}$$

where $\bar{y}_i = \underline{\iota}_T'\underline{y}_i/\sqrt{T}$, $\bar{\underline{z}}_i' = \underline{\iota}_T'Z_i/\sqrt{T}$, $\bar{u}_i = \underline{\iota}_T'\underline{u}_i/\sqrt{T}$, $\tilde{\underline{y}}_i = C_1\underline{y}_i$, $\tilde{Z}_i = C_1Z_i$ and $\tilde{\underline{u}}_i = C_1\underline{u}_i$. The variables $\bar{u}_i$ and $\tilde{\underline{u}}_i$ are uncorrelated because $E\bar{u}_i\tilde{\underline{u}}_i' = \underline{\iota}_T'\Omega C_1'/\sqrt{T} = \underline{0}'$ and

$$E\bar{u}_i = 0 \quad , \quad E\underline{\tilde{u}}_i = 0 \quad , \quad E\bar{u}_i^{\,2} = \sigma^2(1 - \rho + T\rho) \quad ,$$

$$E\underline{\tilde{u}}_i\underline{\tilde{u}}_i' = \sigma^2(1 - \rho)I_{T'} \tag{2.3.6}$$

where $T' = T - 1$.

We can minimize $\underline{\tilde{u}}_i'\underline{\tilde{u}}_i$ to obtain

$$\underline{\tilde{d}}_i = (\tilde{Z}_i'\tilde{Z}_i)^{-1}\tilde{Z}_i'\underline{\tilde{y}}_i$$

$$= [Z_i'(I_T-\underline{\iota}_T\underline{\iota}_T'T^{-1})Z_i]^{-1}Z_i'\,(I_T-\underline{\iota}_T\underline{\iota}_T'T^{-1})\underline{y}_i \tag{2.3.7}$$

as an estimator of $\underline{\delta}$. We can easily recognize that $\underline{\tilde{d}}_i$ is an OLS estimator of $\underline{\delta}$ applied to observations after expressing them as deviations from their respective means. Notice that C_1 is not unique but $C_1'C_1 = I_T - \underline{\iota}_T\underline{\iota}_T'/T$ is unique. The estimator $\underline{\tilde{d}}_i$ depends upon $C_1'C_1$ and hence unique. We can easily show that $E\underline{\tilde{d}}_i = \underline{\delta}$ and the variance-covariance matrix of $\underline{\tilde{d}}_i$ is

$$V(\underline{\tilde{d}}_i) = \sigma^2(1-\rho)(Z_i'Z_i - Z_i'\underline{\iota}_T\underline{\iota}_T'Z_iT^{-1})^{-1}$$

$$= \sigma^2(1-\rho)(Z_i'Z_i)^{-1}[I_{K'} + (\underline{\iota}_T'M_i\underline{\iota}_T)^{-1}Z_i'\underline{\iota}_T\underline{\iota}_T'Z_i\,(Z_i'Z_i)^{-1}] \tag{2.3.8}$$

where $K' = K - 1$, $M_i = I_{K'} - Z_i(Z_i'Z_i)^{-1}Z_i'$ and the second equality is based on the result in Lemma 2.2.2. We can estimate $\sigma^2(1-\rho)$ unbiasedly by

$$(\underline{y}_i - Z_i\underline{\tilde{d}}_i)'\,C_1'C_1(\underline{y}_i - Z_i\underline{\tilde{d}}_i)/(T - K) \quad . \tag{2.3.9}$$

In the transformed equation $\bar{y}_i = \underline{\bar{z}}_i'\underline{\delta} + \bar{u}_i$ there is only one degree of freedom (d.f.) to estimate K-1 regression coefficients and the variance function, $\sigma^2(1-\rho+T\rho)$. We can use this observation to estimate a function of K-1 coefficients. That is, we can take $\bar{y}_i$ as an estimator of $\underline{\bar{z}}_i'\underline{\delta}$ which is estimable.[6] Since $\underline{\tilde{d}}_i$ and

---

[6]A definition of estimable functions is given in Rao (1965a, p. 181).

$\bar{y}_i$ are the estimators of different functions of $\underline{\delta}$, there is no question of pooling them. After estimating the functions of $\underline{\delta}$ and the variance function $\sigma^2(1-\rho)$, we are not left with any d.f. to estimate $\sigma^2(1-\rho+T\rho)$.

Instead of using the observation $\bar{y}_i$ to estimate $\underline{\bar{z}}_i'\underline{\delta}$, we can also use it to estimate $\sigma^2\rho$. Consider the statistic $(\bar{y}_i - \underline{\bar{z}}_i'\underline{\tilde{d}}_i)^2$ and take its expectation to obtain

$$E(\bar{y}_i - \underline{\bar{z}}_i'\underline{\tilde{d}}_i)^2 = E[\bar{u}_i - \underline{\bar{z}}_i'(\tilde{Z}_i'\tilde{Z}_i)^{-1}\tilde{Z}_i'\underline{\tilde{u}}_i]^2$$

$$= \sigma^2(1-\rho)[1 + \underline{\bar{z}}_i'(\tilde{Z}_i'\tilde{Z}_i)^{-1}\underline{\bar{z}}_i] + T\sigma^2\rho \ . \quad (2.3.10)$$

Consequently, an unbiased estimator of $\sigma^2\rho$ is

$$\frac{1}{T}(\bar{y}_i - \underline{\bar{z}}_i'\underline{\tilde{d}}_i)^2 - \frac{1}{T(T-K)}(\underline{\tilde{y}}_i - \tilde{Z}_i\underline{\tilde{d}}_i)'(\underline{\tilde{y}}_i - \tilde{Z}_i\underline{\tilde{d}}_i)[1 + \underline{\bar{z}}_i'(\tilde{Z}_i'\tilde{Z}_i)^{-1}\underline{\bar{z}}_i] \ . \quad (2.3.11)$$

If we divide (2.3.11) by (2.3.9) we obtain an estimator, say $\hat{\xi}$, of $\xi = \rho/(1-\rho)$. The estimator $\underline{d}_i(\rho)$ can be written as

$$\underline{d}_i(\xi) = \left[\tilde{Z}_i'\tilde{Z}_i + \frac{\underline{\bar{z}}_i\underline{\bar{z}}_i'}{(1+T\xi)}\right]^{-1}\left[\tilde{Z}_i'\underline{\tilde{y}}_i + \frac{\underline{\bar{z}}_i\bar{y}_i}{(1+T\xi)}\right] \quad (2.3.12)$$

which is a function of ξ. The estimator $\hat{\xi}$ can be substituted in $\underline{d}_i(\xi)$ to obtain an approximate Aitken estimator, say $\underline{d}_i(\hat{\xi})$, but in the process the observation $\bar{y}_i$ is used twice--once to estimate $\sigma^2\rho$ and again to estimate $\underline{\delta}$. The vectors $\underline{\bar{z}}_i\bar{y}_i$ and $\tilde{Z}_i'\underline{\tilde{y}}_i$ are not independent of $\hat{\xi}$. In a situation like this we can use either $\underline{\tilde{d}}_i$ or the OLS estimator

$$\underline{d}_i = (Z_i'Z_i)^{-1} Z_i'\underline{y}_i \ . \quad (2.3.13)$$

It is easy to show that $E\underline{d}_i = \underline{\delta}$ and the variance-covariance matrix of $\underline{d}_i$ is

$$V(\underline{d}_i) = \sigma^2(Z_i'Z_i)^{-1}Z_i'\Omega Z_i(Z_i'Z_i)^{-1}$$

$$= \sigma^2(1-\rho)(Z_i'Z_i)^{-1}[I_{K'} + \rho(1-\rho)^{-1}Z_i'\underline{\iota}_T\underline{\iota}_T'Z_i(Z_i'Z_i)^{-1}] \ . \quad (2.3.14)$$

We say that the estimator $\tilde{\underline{d}}_i$ is more efficient than the estimator $\underline{d}_i$ if $\underline{\ell}'V(\tilde{\underline{d}}_i)\underline{\ell} < \underline{\ell}'V(\underline{d}_i)\underline{\ell}$ for every (K-1) x 1 nonnull vector $\underline{\ell}$. With the model we are considering [(2.3.1) with no constant term], we prove the following theorems.

Theorem 2.3.1: The OLS estimator $\underline{d}_i$ is more efficient than $\tilde{\underline{d}}_i$ if $-(T-1)^{-1} < \rho < (\underline{\iota}_T'M_i\underline{\iota}_T + 1)^{-1}$, as efficient as $\tilde{\underline{d}}_i$ if $\rho = (\underline{\iota}_T'M_i\underline{\iota}_T+1)^{-1}$, and less efficient than $\tilde{\underline{d}}_i$ if $(\underline{\iota}_T'M_i\underline{\iota}_T+1)^{-1} < \rho < 1$.[7]

Proof: It is obvious that $\underline{\iota}_T'Z_i(Z_i'Z_i)^{-1}Z_i'\underline{\iota}_T > 0$ and $\underline{\ell}'Z_i'\underline{\iota}_T\underline{\iota}_T'Z_i(Z_i'Z_i)^{-1}\underline{\ell} \geq 0$ for any nonnull vector $\underline{\ell}$. Since $\underline{\iota}_T'M_i\underline{\iota}_T$ is an idempotent quadratic form, $\underline{\iota}_T'M_i\underline{\iota}_T \geq 0$. We assume that $\underline{\iota}_T'M_i\underline{\iota}_T \neq 0$. Subtracting (2.3.14) from (2.3.8) we have

$$V(\tilde{\underline{d}}_i) - V(\underline{d}_i) = \sigma^2(1-\rho)(Z_i'Z_i)^{-1}Z_i'\underline{\iota}_T\underline{\iota}_T'Z_i(Z_i'Z_i)^{-1} \cdot [(\underline{\iota}_T'M_i\underline{\iota}_T)^{-1} - \rho(1-\rho)^{-1}] \quad . \tag{2.3.15}$$

The matrix $V(\tilde{\underline{d}}_i) - V(\underline{d}_i)$ is positive semidefinite if $(\underline{\iota}_T'M_i\underline{\iota}_T)^{-1} \geq \rho(1-\rho)^{-1}$. This implies that $\underline{d}_i$ is more efficient than $\tilde{\underline{d}}_i$ if $\rho < (\underline{\iota}_T'M_i\underline{\iota}_T+1)^{-1}$ and as efficient as $\tilde{\underline{d}}_i$ if $\rho = (\underline{\iota}_T'M_i\underline{\iota}_T+1)^{-1}$. The matrix $V(\tilde{\underline{d}}_i) - V(\underline{d}_i)$ is negative definite if $(\underline{\iota}_T'M_i\underline{\iota}_T)^{-1} < \rho(1-\rho)^{-1}$. That is, $\tilde{\underline{d}}_i$ is more efficient than $\underline{d}_i$ if $\rho > (\underline{\iota}_T'M_i\underline{\iota}_T+1)^{-1}$.

Theorem 2.3.2: The Aitken estimator $\underline{d}_i(\rho)$ is the same as $\underline{d}_i$ or $\tilde{\underline{d}}_i$ in any given sample if $\underline{\iota}_T'Z_i = 0$.

Proof: Using the result (ii) of Lemma 2.2.1 and the condition $\underline{\iota}_T'Z_i = 0$, we can easily show that $\underline{d}_i(\rho) = \tilde{\underline{d}}_i = \underline{d}_i$ in any given sample.

Theorem 2.3.3: If $\lim_{T\to\infty} T^{-1}Z_i'Z_i$ is finite and positive definite and $\lim_{T\to\infty} T^{-1}\underline{\iota}_T'Z_i$ is finite and nonnull, the estimator $\tilde{\underline{d}}_i$ is asymptotically as efficient as the Aitken estimator $\underline{d}_i(\rho)$.

---

[7]Theorem 2.3.1 is an extension of Problem 3 in Rao (1965a, p. 249).

Proof: As $T\to\infty$, the interval $-(T-1)^{-1}$ to $(\underline{\iota}_T' M_i \underline{\iota}_T + 1)^{-1}$ degenerates to a point at zero ruling out the possibility that the OLS estimator $\underline{d}_i$ is more efficient than the estimator $\tilde{\underline{d}}_i$. Using the result (ii) in Lemma 2.2.1 we write

$$\underline{d}_i(\rho) = [T^{-1}Z_i'Z_i - T\rho(1-\rho+T\rho)^{-1}T^{-2}Z_i'\underline{\iota}_T\underline{\iota}_T'Z_i]^{-1}[T^{-1}Z_i'\underline{y}_i$$

$$- T\rho(1-\rho+T\rho)^{-1}T^{-2}Z_i'\underline{\iota}_T\underline{\iota}_T'\underline{y}_i] \quad . \qquad (2.3.16)$$

Since $\lim_{T\to\infty} T\rho(1-\rho+T\rho)^{-1} = 1$, $\operatorname{plim}_{T\to\infty}|\tilde{\underline{d}}_i - \underline{d}_i(\rho)| = 0$. According to a limit theorem in Rao (1965a, p. 101) the limiting distribution of $\tilde{\underline{d}}_i$ is the same as that of $\underline{d}_i(\rho)$ which is the BLUE of $\underline{\delta}$. Hence, $\tilde{\underline{d}}_i$ is asymptotically efficient.

2.4 Analysis of Time Series of Cross-Sections

(a) A Class of Estimators of Coefficients

The parameter ρ can be estimated if we have repeated observations on $\underline{y}_i$, cf. Halperin (1951), and Balestra and Nerlove (1966). Suppose that a time series of cross-sections on all the variables in (2.3.1) is available. Combining all the T observations on the ith individual unit we write

$$\underline{y}_i = X_i\underline{\beta} + \underline{u}_i \qquad (i = 1,2,\ldots,n) \quad , \qquad (2.4.1)$$

where the symbols $\underline{y}_i$, X_i, $\underline{\beta}$ and $\underline{u}_i$ are as explained in (2.3.1). There are T observations on each variable for each of n individual units. Following Balestra and Nerlove (1966) we write

$$\underline{u}_i = \underline{\iota}_T\mu_i + \underline{\nu}_i \qquad (i = 1,2,\ldots,n) \quad , \qquad (2.4.2)$$

where μ_i is a time-invariant individual effect and $\underline{\nu}_i \equiv (\nu_{i1},\nu_{i2},\ldots,\nu_{iT})'$ is a Tx1 vector of remainders. We will make use of the following assumption in this section.

Assumption 2.4.1:

(1) The sample sizes n and T are such that $n > K$ and $T > 1$.

(2) The independent variables are nonstochastic in the sense that the X_i is fixed in repeated samples on $\underline{y}_i$. The rank of $X \equiv [X_1', X_2', \ldots, X_n']'$ is K.

(3) The μ_i's are independently and identically distributed with $E\mu_i = 0$ and $E\mu_i^2 = \sigma_\mu^2$.

(4) The $\underline{\nu}_i$'s are independently and identically distributed with $E\underline{\nu}_i = 0$ and $E\underline{\nu}_i\underline{\nu}_i' = \sigma_\nu^2 I_T$.

(5) μ_i and ν_{jt} are independent for all i & j.

The specifications (3), (4), and (5) in Assumption 2.4.1 imply that

$$E\underline{u}_i\underline{u}_j' = \begin{cases} \sigma^2\Omega & \text{if } i = j \\ 0 & \text{otherwise} \end{cases} , \tag{2.4.3}$$

where Ω is as shown in Assumption 2.3.1, $\rho = \sigma_\mu^2/\sigma^2$, $\sigma^2 = \sigma_\mu^2 + \sigma_\nu^2$. Clearly, ρ cannot be negative in this case. We restrict the range of ρ to be $0 \leq \rho < 1$. As we have already pointed out in subsection 1.2(a), the specification (2.4.2) under Assumption 2.4.1 implies that the intercept of the model (2.4.1) varies randomly across units.

Applying Aitken's generalized least squares to (2.4.1) we obtain

$$\underline{b}(\rho) = \left(\sum_{j=1}^{n} X_j'\Omega^{-1}X_j\right)^{-1}\left(\sum_{i=1}^{n} X_i'\,\Omega^{-1}\underline{y}_i\right) . \tag{2.4.4}$$

The estimator $\underline{b}(\rho)$ is the BLUE of $\underline{\beta}$. Since $\underline{\iota}_T$ is the first column of X_i, we have $(X_i'X_i)^{-1}X_i'\underline{\iota}_T = \underline{i}_1$ which is the first column of an identity matrix I_K. Consequently, using the result (ii) in Lemma 2.2.1 we have

$$\underline{b}(\rho) = \left[\sum_{j=1}^{n} X_j'X_j\left\{I_K - \rho(1-\rho+T\rho)^{-1}\underline{i}_1\underline{\iota}_T'X_j\right\}\right]^{-1}$$

$$\cdot\left[\sum_{\mu=1}^{n} X_\mu'X_\mu\left\{I_K - \rho(1-\rho+T\rho)^{-1}\underline{i}_1\underline{\iota}_T'X_\mu\right\}\underline{b}_\mu\right] , \tag{2.4.4a}$$

where $\underline{b}_{\mu} = (X_{\mu}'X_{\mu})^{-1}X_{\mu}'\underline{y}_{\mu}$. When $\rho = 0$ (or $\underline{\iota}_T'X_i = 0$ for every i), $\underline{b}(\rho)$ reduces to the OLS estimator,

$$\underline{b} = \left(\sum_{j=1}^{n} X_j'X_j\right)^{-1}\left(\sum_{i=1}^{n} X_i'X_i\underline{b}_i\right) . \tag{2.4.5}$$

If we compare (2.4.4a) with (2.4.5) we can recognize that both the Aitken and OLS estimators are the weighted averages of $\underline{b}_i$'s but the weight given to $\underline{b}_i$ in (2.4.4a) is different from that given to the same $\underline{b}_i$ in (2.4.5). When $\rho \neq 0$ and is unknown (or $\underline{\iota}_T'X_i \neq 0$ for every i), we cannot use the estimator $\underline{b}(\rho)$. In this case we can consider an Aitken estimator based on an estimator of ρ. Now it will be convenient to consider the following transformation of observations.

Using the orthogonal matrix 0_T which diagonalizes Ω, we can transform the observations in (2.4.1) as

$$\bar{\underline{y}} = \bar{X}\underline{\beta} + \bar{\underline{u}} \tag{2.4.6a}$$

$$\tilde{\underline{y}} = \tilde{Z}\underline{\delta} + \tilde{\underline{u}} \quad , \tag{2.4.6b}$$

where $\bar{\underline{y}} = (I_n \otimes \underline{\iota}_T'/\sqrt{T})\underline{y}$ is a n x 1 vector of transformed observations on the dependent variable, $\otimes$ denotes the Kronecker product, $\underline{y} \equiv [\underline{y}_1',\underline{y}_2',\ldots,\underline{y}_n']'$, $\bar{X} \equiv (I_n \otimes \underline{\iota}_T'/\sqrt{T})X$ is a nXK matrix of transformed observations on the independent variables, $X \equiv [X_1',X_2',\ldots,X_n']'$, $\bar{\underline{u}} \equiv (I_n \otimes \underline{\iota}_T'/\sqrt{T})\underline{u}$ is a n x 1 vector of transformed disturbances, $\underline{u} \equiv [\underline{u}_1',\underline{u}_2',\ldots,\underline{u}_n']'$, $\tilde{\underline{y}} = (I_n \otimes C_1)\underline{y}$ is a n(T-1) x 1 vector of transformed observations on the dependent variable, $\tilde{Z} = (I_n \otimes C_1)Z$ is a n(T-1) x (K-1) matrix of transformed observations on the independent variables, $Z \equiv [Z_1',Z_2',\ldots,Z_n']'$, and $\tilde{\underline{u}} = (I_n \otimes C_1)\underline{u}$ is a n(T-1) x 1 vector of transformed disturbances. In (2.4.6b) use is made of the result $(I_n \otimes C_1)\underline{\iota}_{nT} = 0$. We can easily recognize that $1/\sqrt{T}$ times each observation in (2.4.6a) is the mean of time series observations on a variable for a unit. It follows from (2.4.3) and the properties of $\underline{\iota}_T$ and C_1 that

$$E\bar{\underline{u}} = 0 \ , \quad E\tilde{\underline{u}} = 0 \ , \quad E\bar{\underline{u}}\,\bar{\underline{u}}' = \bar{\sigma}^2 I_n \ , \quad \text{and } E\tilde{\underline{u}}\,\tilde{\underline{u}}' = \sigma_{\nu}^2 I_{nT'} \ , \tag{2.4.7}$$

where $\bar{\sigma}^2 = \sigma^2(1-\rho+T\rho) = \sigma_\nu^2 + T\sigma_\mu^2$ and $\sigma_\nu^2 = \sigma^2(1-\rho)$. Notice that the parameter $\bar{\sigma}^2$ depends upon T. In the reparameterized version of the model the restriction $\bar{\sigma}^2 \geq \sigma_\nu^2$ is present.[8] If we combine the information about common slope coefficients in (2.4.6a) and (2.4.6b) we obtain a model with heteroskedastic disturbances. In order to construct an Aitken estimator of $\underline{\delta}$ based on estimated variances of disturbances we first find the estimates of variances. An unbiased estimator of $\bar{\sigma}^2$ is given by

$$\hat{\bar{\sigma}}^2 = \frac{\underline{\bar{y}}'\bar{M}\,\underline{\bar{y}}}{n-K} , \qquad (2.4.8)$$

where $\bar{M} = I_n - \bar{X}(\bar{X}'\bar{X})^{-1}\bar{X}'$. Notice that the matrix $\bar{X}'\bar{X}$ is likely to be nonsingular if $n > K$.[9] An unbiased estimator of σ_ν^2 is given by

$$\hat{\sigma}_\nu^2 = \frac{\underline{\tilde{y}}'\tilde{M}\,\underline{\tilde{y}}}{n(T-1)-K+1} , \qquad (2.4.9)$$

where $\tilde{M} = I_{nT'} - \tilde{Z}(\tilde{Z}'\tilde{Z})^{-1}\tilde{Z}'$. The matrix $\tilde{Z}'\tilde{Z}$ can be nonsingular if $n > K$ and $T > 1$.[10] Using (2.4.8) and (2.4.9) we can form an estimator of ρ as

$$\hat{\rho} = \frac{\hat{\bar{\sigma}}^2 - \hat{\sigma}_\nu^2}{\hat{\bar{\sigma}}^2 + (T-1)\hat{\sigma}_\nu^2} . \qquad (2.4.10)$$

Notice that ρ is estimable only if $n > K$ and $T > 1$. The structure of the estimator $\hat{\rho}$ is such that it can assume negative values with positive probability.

---

[8] This restriction will not be present if we specify the model in such a way that the pairwise correlation between the elements of $\underline{u}_i$ is equal to ρ and $-(T-1)^{-1} < \rho < 1$. In this case $\sigma^2(1-\rho+T\rho) < \sigma^2(1-\rho)$ if $-(T-1)^{-1} < \rho < 0$ and $\sigma^2(1-\rho+T\rho) \geq \sigma^2(1-\rho)$ if $0 \leq \rho < 1$.

[9] If $\bar{X}'\bar{X}$ is singular even when $n > K$, we can take $\bar{M} = I_n - \bar{X}(\bar{X}'\bar{X})^- \bar{X}'$, where $(\bar{X}'\bar{X})^-$ is any generalized inverse of $\bar{X}'\bar{X}$, cf. Rao (1965a, pp. 185-186). Furthermore, n-K, wherever occurs, is to be replaced by n-rank $(\bar{X})$.

[10] If $\tilde{Z}'\tilde{Z}$ is singular even when $n > K$ and $T > 1$, we can take $\tilde{M} = I_{nT'} - \tilde{Z}(\tilde{Z}'\tilde{Z})^-\tilde{Z}'$ and replace $n(T-1)-K+1$, wherever occurs, by $n(T-1)$-rank$(\tilde{X})$.

This may be the direct consequence of neglecting prior information on the ranges of coefficients and the restriction $\bar{\sigma}^2 \geq \sigma_\nu^2$ implied by the model. However, we can use our procedure to detect a possible misspecification in (2.4.1) - (2.4.3). If ρ assumes a large negative value in any application it may mean either that the basic model is misspecified or that the sample data cannot discriminate between σ_ν^2 and σ_μ^2. This point will be discussed further in the Appendix to this chapter.

Pooling the information in (2.4.6a) and (2.4.6b) about common coefficients $\underline{\delta}$ we can consider $\underline{\alpha}$-I class estimators of the following form.

$$\underline{d}(\hat{\rho},\underline{\alpha}) = \left[\frac{\alpha_0\bar{Z}'N_1\bar{Z}}{\alpha_1\hat{\bar{\sigma}}^2 + \alpha_2\hat{\sigma}_\nu^2} + \frac{\alpha_3\tilde{Z}'\tilde{Z}}{\hat{\sigma}_\nu^2}\right]^{-1}\left[\frac{\alpha_0\bar{Z}'N_1\bar{\underline{y}}}{\alpha_1\hat{\bar{\sigma}}^2 + \alpha_2\hat{\sigma}_\nu^2} + \frac{\alpha_3\tilde{Z}'\tilde{\underline{y}}}{\hat{\sigma}_\nu^2}\right] \quad (2.4.11)$$

where $\bar{X} \equiv [\sqrt{T}\underline{\iota}_n, \bar{Z}]$, $\bar{Z} \equiv (I_n \otimes \underline{\iota}_T'/\sqrt{T})Z$, $N_1 = I_n - \underline{\iota}_n\underline{\iota}_n'/n$, $\underline{\iota}_n$ is a n x 1 vector of unit elements and $\underline{\alpha} = (\alpha_0,\alpha_1,\alpha_2,\alpha_3)'$ is a 4 x 1 vector of nonnegative constants obeying the restriction that α_1 and α_2 cannot be equal to zero simultaneously.[11]

An Aitken estimator of $\underline{\delta}$ based on $\hat{\rho}$, which we denote by $\underline{d}(\hat{\rho})$, is the $\underline{\alpha}$-I estimator with $\underline{\alpha} = (1,1,0,1)'$ or with ($\alpha_0 = \alpha_1, \alpha_2 = 0$ and $\alpha_3 = 1$) or with ($\alpha_0 = \alpha_3 \neq 0$, $\alpha_1 = 1$, $\alpha_2 = 0$). We can write

$$\underline{d}(\hat{\rho}) = (\hat{V}^{-1} + \tilde{V}^{-1})^{-1}(\hat{V}^{-1}\,\bar{\underline{d}} + \tilde{V}^{-1}\,\tilde{\underline{d}}) \quad (2.4.12)$$

where $\bar{\underline{d}} = (\bar{Z}'N_1\bar{Z})^{-1}\,\bar{Z}'N_1\bar{\underline{y}}$, $\hat{V} = \hat{\bar{\sigma}}^2(\bar{Z}'N_1\bar{Z})^{-1}$, $\tilde{\underline{d}} = (\tilde{Z}'\tilde{Z})^{-1}\,\tilde{Z}'\tilde{\underline{y}}$ and $\tilde{V} = \hat{\sigma}_\nu^2(\tilde{Z}'\tilde{Z})^{-1}$. This shows that $\underline{d}(\hat{\rho})$ is a weighted average of the OLS estimators $\bar{\underline{d}}$ and $\tilde{\underline{d}}$, the weights being the inverses of estimated variance-covariance matrices $\hat{V}$ and $\tilde{V}$. $\bar{\underline{d}}$ and $\tilde{\underline{d}}$ are the least squares quantities obtained from the least squares fits of $N_1\bar{\underline{y}}$ on $N_1\bar{Z}$ and $\tilde{\underline{y}}$ on $\tilde{Z}$, respectively. $\bar{\underline{d}}$ is the $\underline{\alpha}$-I estimator with $\underline{\alpha} = (\alpha_0 \neq 0, \alpha_1, \alpha_2, 0)'$ and $\tilde{\underline{d}}$ is the $\underline{\alpha}$-I estimator with $\underline{\alpha} = (0, \alpha_1, \alpha_2, \alpha_3 \neq 0)'$. Pooling the information in (2.4.6a) and (2.4.6b) about $\underline{\delta}$ and applying OLS we obtain

---

[11] If $\bar{Z}'N_1\bar{Z}$ and $\tilde{Z}'\tilde{Z}$ are singular, the estimator $\underline{d}(\rho,\underline{\alpha})$ exists only when rank $(N_1\bar{Z})$ = $K-m_1-1$, $(m_1 < K-1)$, rank $(\tilde{Z}) = m_1$ such that $C(\bar{Z}'N_1) \cap C(\tilde{Z}')$ is null, where $C(\tilde{Z}')$ indicates the linear subspace spanned by the columns of $\tilde{Z}$.

$$\underline{d} = (\bar{Z}'N_1\bar{Z} + \tilde{Z}'\tilde{Z})^{-1}(\bar{Z}'N_1\bar{\underline{y}} + \tilde{Z}'\tilde{\underline{y}}) \qquad . \qquad (2.4.13)$$

We can easily recognize that $\underline{d}$ is the $\underline{\alpha}$-I estimator with $\underline{\alpha} = (1,0,1,1)'$ or with $(\alpha_1=0, \alpha_2=1, \alpha_0 = \alpha_3 \neq 0)$ or with $(\alpha_0 = \alpha_2, \alpha_1 = 0, \alpha_3 = 1)$. The constants α_0 and α_3 simply change the relative weight given to $\bar{\underline{d}}$ and $\tilde{\underline{d}}$ in $\underline{d}(\hat{\rho},\underline{\alpha})$. An estimator of β_0 is given by

$$b_0(\hat{\rho},\underline{\alpha}) = (n\sqrt{T})^{-1} \sum_{i=1}^{n} [\bar{y}_i - \bar{\underline{z}}_i' \, \underline{d}(\hat{\rho},\underline{\alpha})] \qquad , \qquad (2.4.14)$$

where $\bar{\underline{z}}_i' = \underline{\iota}_T' Z_i/\sqrt{T}$. Let $\underline{b}(\hat{\rho},\underline{\alpha}) = [b_0(\hat{\rho},\underline{\alpha}),\underline{d}'(\hat{\rho},\underline{\alpha})]'$.

The values for the α_i's are to be chosen such that the generalized variance of $\underline{b}(\hat{\rho},\underline{\alpha})$ is minimum or if possible, the variance of each element of $\underline{b}(\hat{\rho},\underline{\alpha})$ is minimum.

(b) Computational Procedure

It will be convenient for computational purposes to rewrite (2.4.6a) as

$$T^{-\frac{1}{2}} \bar{\underline{y}} = T^{-\frac{1}{2}} \bar{X}\underline{\beta} + T^{-\frac{1}{2}} \bar{\underline{u}} \qquad .$$

Now the above equation is in terms of the means of time series observations on different variables for different units. We can write

$$T^{-1} \bar{Z}'N_1\bar{Z} = \left[\sum_{i=1}^{n} \frac{Z_i'\underline{\iota}_T}{T} \left(\frac{\underline{\iota}_T'Z_i}{T} - \frac{1}{n}\sum_{i=1}^{n} \frac{\underline{\iota}_T'Z_i}{T}\right)\right]$$

$$= \left[\sum_{i=1}^{n} \left(\frac{Z_i'\underline{\iota}_T}{T} - \frac{1}{n}\sum_{i=1}^{n} \frac{Z_i'\underline{\iota}_T}{T}\right) \left(\frac{\underline{\iota}_T'Z_i}{T} - \frac{1}{n}\sum_{i=1}^{n} \frac{\underline{\iota}_T'Z_i}{T}\right)\right] \qquad .$$

Similarly, we can write

$$T^{-1} \bar{Z}'N_1\bar{\underline{y}} = \left[\sum_{i=1}^{n} \frac{Z_i'\underline{\iota}_T}{T} \left(\frac{\underline{\iota}_T'\underline{y}_i}{T} - \frac{1}{n}\sum_{i=1}^{n} \frac{\underline{\iota}_T'\underline{y}_i}{T}\right)\right]$$

$$= \left[\sum_{i=1}^{n} \left(\frac{Z_i'\underline{\iota}_T}{T} - \frac{1}{n}\sum_{i=1}^{n} \frac{Z_i'\underline{\iota}_T}{T}\right) \left(\frac{\underline{\iota}_T'\underline{y}_i}{T} - \frac{1}{n}\sum_{i=1}^{n} \frac{\underline{\iota}_T'\underline{y}_i}{T}\right)\right] \qquad .$$

Thus, $(\bar{Z}'N_1\bar{Z})^{-1}\bar{Z}'N_1\bar{\underline{y}}$ is a least squares quantity obtained by applying OLS to means of time series observations after expressing them as deviations from their respective means. The estimator of $\bar{\sigma}^2$ in (2.4.8) can be obtained by taking the sum of squares of residuals from a least squares fit of $T^{-\frac{1}{2}}\bar{\underline{y}}$ upon $T^{-\frac{1}{2}}\bar{X}$, and dividing it by $T^{-1}(n-K)$. Remembering that $C_1'C_1 = I_T - \underline{\iota}_T\underline{\iota}_T'/T$ we can write

$$\tilde{Z}'\tilde{Z} = \sum_{i=1}^{n} Z_i'(I_T - T^{-1}\underline{\iota}_T\underline{\iota}_T')Z_i \quad \text{and} \quad \tilde{Z}'\tilde{\underline{y}} = \sum_{i=1}^{n} Z_i'(I_T - T^{-1}\underline{\iota}_T\underline{\iota}_T')\underline{y}_i \quad .$$

Thus, $\tilde{Z}'\tilde{Z}$ is the matrix of sum of squares and products of observations on the independent variables expressed as deviations from their respective means of time series observations. We can first express the T time series observations on each variable for each unit in (2.4.1a) as deviations from their own mean and then apply OLS to obtain $\tilde{\underline{d}}$. The estimator of σ_ν^2 in (2.4.9) is obtained by taking the sum of squares of residuals from this least squares fit and dividing it by n(T-1)-K+1. For computational purposes the matrix C_1 is not important. It is only useful for obtaining analytical results.

(c) <u>The Exact Finite Sample Properties of $\underline{d}(\hat{\rho},\underline{\alpha})$</u>

In this subsection we make use of the following assumption.

<u>Assumption 2.4.2</u>: The disturbance vector $\underline{u}$ is normally distributed.

Substituting (2.4.6a) and (2.4.6b) into (2.4.11) we have

$$\underline{d}(\hat{\rho},\underline{\alpha}) = \underline{\delta} + \left[\frac{\alpha_0\bar{Z}'N_1\bar{Z}}{\alpha_1\hat{\bar{\sigma}}^2 + \alpha_2\hat{\sigma}_\nu^2} + \frac{\alpha_3\tilde{Z}'\tilde{Z}}{\hat{\sigma}_\nu^2}\right]^{-1}\left[\frac{\alpha_0\bar{Z}'N_1\bar{\underline{u}}}{\alpha_1\hat{\bar{\sigma}}^2 + \alpha_2\hat{\sigma}_\nu^2} + \frac{\alpha_3\tilde{Z}'\tilde{\underline{u}}}{\hat{\sigma}_\nu^2}\right] \quad . \qquad (2.4.15)$$

Due to Assumption 2.4.2, $\bar{u}_i = \underline{\iota}_T'\underline{u}_i/\sqrt{T}$ is independent of $\tilde{\underline{u}}_j = C_1\underline{u}_j$ for every i & j. Therefore, $\bar{\underline{u}}$ is independent of $\tilde{\underline{u}}$. Since $\bar{M}\bar{X} = 0$ and $\tilde{M}\tilde{Z} = 0$, $\bar{Z}'N_1\bar{\underline{u}}$ is independent of $\hat{\bar{\sigma}}^2$ and $\tilde{Z}'\tilde{\underline{u}}$ is independent of $\hat{\sigma}_\nu^2$. Now from the definition of conditional expectations [cf. Rao (1965a, p. 79)],

$$E[\underline{d}(\hat{\rho},\underline{\alpha})] = E_{\hat{\rho}}E[\underline{d}(\hat{\rho},\underline{\alpha})|\hat{\rho}] = E_{\hat{\rho}}(\underline{\delta}) = \underline{\delta} \quad . \tag{2.4.16}$$

This implies that $b_0(\hat{\rho},\underline{\alpha})$ in (2.4.14) is also unbiased. The variance-covariance matrix of $\underline{d}(\hat{\rho},\underline{\alpha})$ conditional on $\hat{\bar{\sigma}}^2$ and $\hat{\sigma}_\nu^2$ is

$$\left[\frac{\alpha_0\bar{Z}'N_1\bar{Z}}{\alpha_1\hat{\bar{\sigma}}^2 + \alpha_2\hat{\sigma}_\nu^2} + \frac{\alpha_3\tilde{Z}'\tilde{Z}}{\hat{\sigma}_\nu^2}\right]^{-1}\left[\frac{\alpha_0^2\bar{\sigma}^2\bar{Z}'N_1\bar{Z}}{(\alpha_1\hat{\bar{\sigma}}^2 + \alpha_2\hat{\sigma}_\nu^2)^2} + \frac{\alpha_3^2\sigma_\nu^2\tilde{Z}'\tilde{Z}}{\hat{\sigma}_\nu^4}\right]\left[\frac{\alpha_0\bar{Z}'N_1\bar{Z}}{\alpha_1\hat{\bar{\sigma}}^2 + \alpha_2\hat{\sigma}_\nu^2} + \frac{\alpha_3\tilde{Z}'\tilde{Z}}{\hat{\sigma}_\nu^2}\right]^{-1} . \tag{2.4.17}$$

The unconditional variance-covariance matrix of $\underline{d}(\hat{\rho},\underline{\alpha})$ is simply the expectation of (2.4.17) over the distributions of $\hat{\bar{\sigma}}^2$ and $\hat{\sigma}_\nu^2$.

Since $\bar{Z}'N_1\bar{Z}$ is assumed to be positive definite, it follows from Lemma 2.2.3 that there exists a nonsingular matrix P of order K-1 such that $P'\bar{Z}'N_1\bar{Z}P = I_{K'}$ and $P'\tilde{Z}'\tilde{Z}P = \Lambda_{K'}$, a diagonal matrix.[12] Let λ_k be the kth diagonal element of $\Lambda_{K'}$. Since $\tilde{Z}'\tilde{Z}$ is assumed to be positive definite, all the diagonal elements of $\Lambda_{K'}$ are positive. The λ_k's are the roots of $|\tilde{Z}'\tilde{Z} - \lambda\bar{Z}'N_1\bar{Z}| = 0$. Using P we can write the variance-covariance matrix of $\underline{d}(\hat{\rho},\underline{\alpha})$ as

$$V[\underline{d}(\hat{\rho},\underline{\alpha})] = \sigma_\nu^2 \, PE_fD(f)P' \quad , \tag{2.4.18}$$

where D(f) is a diagonal matrix with $\left[\alpha_0^2c + \alpha_3^2\lambda_k(\alpha_1f + \alpha_2)^2\right]\left[\alpha_0 + \alpha_3\lambda_k(\alpha_1f + \alpha_2)\right]^{-2}$ as its kth diagonal element, $c = \bar{\sigma}^2/\sigma_\nu^2$, $f = \hat{\bar{\sigma}}^2/\hat{\sigma}_\nu^2$, and E_f stands for the expectation over the distribution of f. It is shown in the Appendix to this chapter that the expectation of each diagonal element of D(f) is finite. Notice that the constant α_2/α_1 has the effect of changing the origin of the f-distribution and the constants λ_k, α_1, and α_3 have the effect of changing the scale of f-distribution.

It follows from (2.4.18) that the variance-covariance matrix of the estimator $\underline{d}(\hat{\rho})$ is

---

[12] Even when $\bar{Z}'N_1\bar{Z}$ and $\tilde{Z}'\tilde{Z}$ are both singular we can find a matrix P such that $P'\bar{Z}N_1\bar{Z}P$ and $P'\tilde{Z}'\tilde{Z}P$ are both diagonal, cf. Mitra and Rao (1968a). We can easily make necessary changes in our subsequent analysis if $\bar{Z}'N_1\bar{Z}$ and $\tilde{Z}'\tilde{Z}$ are singular and $(\bar{Z}'N_1\bar{Z} + \tilde{Z}'\tilde{Z})$ is nonsingular.

$$V[\underline{d}(\hat{\rho})] = \sigma_{\nu}^2 \, PE_f D_1(f)P' \quad , \tag{2.4.19}$$

where $D_1(f)$ is a diagonal matrix with $(c + \lambda_k f^2)(1 + \lambda_k f)^{-2}$ as its kth diagonal element.

The variance-covariance matrix of $\underline{d}$ in (2.4.13) is

$$\begin{aligned} V(\underline{d}) &= (\overline{Z}'N_1\overline{Z} + \tilde{Z}'\tilde{Z})^{-1}(\overline{\sigma}^2\overline{Z}'N_1\overline{Z} + \sigma_{\nu}^2 \, \tilde{Z}'\tilde{Z})(\overline{Z}'N_1\overline{Z} + \tilde{Z}'\tilde{Z})^{-1} \\ &= \sigma_{\nu}^2 \, PD_2P' \quad , \end{aligned} \tag{2.4.20}$$

where D_2 is a diagonal matrix with $(c + \lambda_k)(1 + \lambda_k)^{-2}$ as its kth diagonal element. If $V(\underline{d})$ and $V[\underline{d}(\hat{\rho})]$ satisfy the condition that

$$\underline{\ell}'(V(\underline{d}) - V[\underline{d}(\hat{\rho})])\underline{\ell} > 0 \tag{2.4.21}$$

for every nonnull K'x1 vector $\underline{\ell}$, then we say that the estimator $\underline{d}(\hat{\rho})$ is more efficient than the estimator $\underline{d}$. In other words when (2.4.21) holds, we say that $\underline{d}(\hat{\rho})$ is superior to $\underline{d}$ according to the minimum variance criterion. We can write (2.4.21) as

$$Q_v = \frac{\underline{\ell}'PE_f D_1(f)P'\underline{\ell}}{\underline{\ell}PD_2P'\underline{\ell}} < 1 \quad . \tag{2.4.22}$$

It follows from Lemma 2.2.5 that

$$\sup_{\underline{\ell}} Q_v = \gamma_{max} \quad , \tag{2.4.23}$$

where γ_{max} is the largest root of $|PE_f D_1(f)P' - \gamma PD_2P'| = 0$. Because P and D_2 are nonsingular matrices, γ_{max} is the largest root of $|E_f D_1(f)D_2^{-1} - \gamma I| = 0$. We also know that $E_f D_1(f)D_2^{-1}$ is a diagonal matrix. Consequently, its characteristic roots are its diagonal elements, cf. Goldberger (1964, p. 29). The largest diagonal element of $E_f D_1(f)D_2^{-1}$ is its largest characteristic root. It is easy to show that the condition (2.4.21) is satisfied if and only if

$$e_{1m} = E_f \frac{(1 + T\xi + \lambda_k f^2)(1 + \lambda_k)^2}{(1 + \lambda_k f)^2 (1 + T\xi + \lambda_k)} < 1 \quad , \qquad (2.4.24)$$

where $\xi = \sigma_\mu^2/\sigma_\nu^2 = \rho(1-\rho)^{-1}$, $c = 1+T\xi$, λ_k is a diagonal element of $\Lambda_{K'}$, and e_{1m} is the largest diagonal elemnt of $E_f D_1(f) D_2^{-1}$, cf. Rao [1965a, p. 56, Problem 9(i)]. e_{1m} is a function of n,T,K,ρ and λ_k. λ_k itself depends upon n and T because it is one of the roots of the determinantal equation $|\tilde{Z}'\tilde{Z} - \lambda\bar{Z}'N_1\bar{Z}| = 0$. The matrix $(\tilde{Z}'\tilde{Z} - \bar{Z}'N_1\bar{Z})$ can be either positive definite or negative definite. It, therefore, follows from Lemma 2.2.4 that the range of λ_k is from 0 to ∞.

The variance-covariance matrix of $\tilde{\underline{d}}$ is

$$V(\tilde{\underline{d}}) = \sigma^2(1-\rho)(\tilde{Z}'\tilde{Z})^{-1}$$

$$= \sigma_\nu^2 \, P\Lambda_{K'}^{-1} P' \quad . \qquad (2.4.25)$$

Again according to the minimum variance criterion the estimator $\underline{d}(\hat{\rho})$ is more efficient than $\tilde{\underline{d}}$ if

$$e_{2m} = E_f \frac{(1 + T\xi + \lambda_k f^2)\lambda_k}{(1 + \lambda_k f)^2} < 1 \quad , \qquad (2.4.26)$$

where e_{2m} is the largest diagonal element of $E_f D_1(f)\Lambda_{K'}$. Furthermore, the estimator $\underline{d}$ is more efficient than $\tilde{\underline{d}}$ if

$$e_{3m} = \frac{(1 + T\xi + \lambda_k)\lambda_k}{(1 + \lambda_k)^2} < 1 \quad , \qquad (2.4.27)$$

where e_{3m} is the largest diagonal element of $D_2\Lambda_{K'}$.

Notice that the sample size n does not explicitly appear in e_{3m} but it affects e_{3m} through λ_k. We evaluated e_{jm} (j = 1,2,3) using numerical integration methods for different sets of values of n,T,K,ρ and λ_k. These are presented in Table 2.1.

The values in Table 2.1 indicate that when the true value of ρ and the sample size T are small, even when n is of moderate size, the OLS estimator $\underline{d}$ is

Table 2.1

Efficiencies of the Estimators of Slope Coefficients

		T=3	T=3, n=7, K=3		T=3, n=11, K=3		T=3, n=19, K=3	
λ_k	ρ	e_{3m}	e_{1m}	e_{2m}	e_{1m}	e_{2m}	e_{1m}	e_{2m}
0.1	0.1	0.118	1.071	0.127	1.033	0.122	1.013	0.120
	0.3	0.197	1.036	0.204	0.994	0.196	0.970	0.191
	0.5	0.339	0.942	0.319	0.899	0.305	0.873	0.296
	0.7	0.669	0.757	0.507	0.716	0.480	0.692	0.463
	0.9	2.322	0.375	0.872	0.344	0.799	0.330	0.766
0.5	0.1	0.407	1.113	0.453	1.057	0.431	1.022	0.416
	0.3	0.619	0.993	0.615	0.933	0.578	0.899	0.557
	0.5	1.000	0.782	0.782	0.724	0.724	0.695	0.695
	0.7	1.889	0.504	0.953	0.457	0.864	0.438	0.828
	0.9	6.333	0.174	1.100	0.155	0.979	0.150	0.949
0.8	0.1	0.527	1.127	0.594	1.061	0.559	1.023	0.539
	0.3	0.762	0.992	0.756	0.921	0.702	0.885	0.674
	0.5	1.185	0.763	0.904	0.696	0.825	0.667	0.791
	0.7	2.173	0.475	1.033	0.426	0.926	0.409	0.889
	0.9	7.111	0.156	1.109	0.140	0.992	0.136	0.969
1.0	0.1	0.583	1.135	0.662	1.063	0.620	1.023	0.596
	0.3	0.821	0.997	0.819	0.919	0.755	0.882	0.724
	0.5	1.250	0.762	0.953	0.691	0.864	0.662	0.828
	0.7	2.250	0.471	1.060	0.421	0.946	0.405	0.911
	0.9	7.250	0.153	1.108	0.137	0.996	0.134	0.975

Table 2.1 (Continued)

λ_k	ρ	T=3 e_{3m}	T=3, n=7, K=3 e_{1m}	e_{2m}	T=3, n=11, K=3 e_{1m}	e_{2m}	T=3, n=19, K=3 e_{1m}	e_{2m}
1.5	0.1	0.680	1.150	0.782	1.065	0.724	1.022	0.695
	0.3	0.908	1.013	0.920	0.922	0.838	0.884	0.803
	0.5	1.320	0.776	1.024	0.696	0.919	0.668	0.882
	0.7	2.280	0.479	1.092	0.426	0.972	0.413	0.941
	0.9	7.080	0.155	1.100	0.141	0.999	0.139	0.983
3.0	0.1	0.812	1.173	0.953	1.063	0.864	1.019	0.828
	0.3	0.991	1.051	1.042	0.941	0.933	0.904	0.896
	0.5	1.313	0.832	1.092	0.741	0.972	0.717	0.941
	0.7	2.063	0.537	1.109	0.482	0.994	0.471	0.971
	0.9	5.813	0.185	1.078	0.172	1.001	0.171	0.992

λ_k	ρ	T=7 e_{3m}	T=7, n=7, K=3 e_{1m}	e_{2m}	T=7, n=11, K=3 e_{1m}	e_{2m}	T=7, n=19, K=3 e_{1m}	e_{2m}
0.1	0.1	0.155	1.033	0.160	1.006	0.156	0.990	0.154
	0.3	0.339	0.921	0.312	0.887	0.301	0.867	0.294
	0.5	0.669	0.745	0.498	0.708	0.474	0.687	0.460
	0.7	1.441	0.508	0.733	0.473	0.682	0.456	0.657
	0.9	5.297	0.194	1.026	0.174	0.920	0.167	0.884
0.5	0.1	0.506	1.047	0.530	0.993	0.502	0.962	0.487
	0.3	1.000	0.774	0.774	0.717	0.717	0.690	0.690
	0.5	1.889	0.500	0.945	0.453	0.857	0.436	0.823
	0.7	3.963	0.267	1.060	0.239	0.946	0.231	0.914
	0.9	14.333	0.076	1.097	0.069	0.994	0.068	0.976

Table 2.1 (Continued)

λ_k	ρ	T=7	T=7, n=7, K=3		T=7, n=11, K=3		T=7, n=19, K=3	
		e_{3m}	e_{1m}	e_{2m}	e_{1m}	e_{2m}	e_{1m}	e_{2m}
0.8	0.1	0.636	1.058	0.673	0.990	0.630	0.956	0.608
	0.3	1.185	0.756	0.896	0.690	0.818	0.663	0.786
	0.5	2.173	0.472	1.025	0.423	0.919	0.407	0.885
	0.7	4.477	0.244	1.091	0.217	0.973	0.211	0.946
	0.9	16.000	0.068	1.087	0.062	0.998	0.061	0.985
1.0	0.1	0.694	1.066	0.740	0.991	0.688	0.955	0.663
	0.3	1.250	0.756	0.945	0.685	0.857	0.659	0.823
	0.5	2.250	0.467	1.052	0.418	0.940	0.403	0.907
	0.7	4.583	0.239	1.097	0.214	0.981	0.209	0.957
	0.9	16.250	0.066	1.080	0.061	0.999	0.061	0.988
1.5	0.1	0.787	1.083	0.852	0.994	0.782	0.955	0.752
	0.3	1.320	0.770	1.016	0.691	0.912	0.665	0.878
	0.5	2.280	0.475	1.084	0.424	0.967	0.411	0.938
	0.7	4.520	0.243	1.100	0.219	0.991	0.215	0.971
	0.9	15.720	0.068	1.067	0.064	1.000	0.063	0.992
3.0	0.1	0.896	1.114	0.998	1.002	0.898	0.964	0.864
	0.3	1.312	0.826	1.084	0.736	0.967	0.714	0.938
	0.5	2.062	0.533	1.100	0.480	0.989	0.470	0.969
	0.7	3.812	0.285	1.086	0.262	0.998	0.258	0.986
	0.9	12.562	0.083	1.046	0.080	1.000	0.079	0.996

Table 2.1 (Continued)

		T=15	T=15, n=7, K=3		T=15, N=11, K=3	
λ_k	ρ	e_{3m}	e_{1m}	e_{2m}	e_{1m}	e_{2m}
0.1	0.1	0.229	0.987	0.226	0.957	0.219
	0.3	0.622	0.763	0.475	0.727	0.452
	0.5	1.331	0.531	0.707	0.495	0.659
	0.7	2.983	0.309	0.921	0.281	0.837
	0.9	11.248	0.095	1.071	0.086	0.966
0.5	0.1	0.704	0.926	0.652	0.869	0.611
	0.3	1.762	0.527	0.928	0.478	0.842
	0.5	3.667	0.286	1.049	.0.256	0.938
	0.7	8.111	0.135	1.096	0.121	0.981
	0.9	30.333	0.035	1.067	0.033	0.997
0.8	0.1	0.856	0.923	0.790	0.853	0.730
	0.3	2.032	0.498	1.012	0.447	0.909
	0.5	4.148	0.261	1.085	0.233	0.968
	0.7	9.086	0.121	1.097	0.109	0.991
	0.9	33.778	0.031	1.054	0.029	0.998
1.0	0.1	0.917	0.927	0.850	0.850	0.780
	0.3	2.107	0.494	1.041	0.442	0.932
	0.5	4.250	0.257	1.093	0.230	0.977
	0.7	9.250	0.118	1.094	0.107	0.994
	0.9	34.250	0.031	1.049	0.029	0.999

Table 2.1 (Continued)

		T=15	T=15, n=7, K=3		T=15, n=11, K=3	
λ_k	ρ	e_{3m}	e_{1m}	e_{2m}	e_{1m}	e_{2m}
1.5	0.1	1.000	0.943	0.943	0.855	0.855
	0.3	2.143	0.503	1.077	0.448	0.961
	0.5	4.200	0.261	1.098	0.235	0.988
	0.7	9.000	0.120	1.084	0.111	0.997
	0.9	33.000	0.031	1.039	0.030	0.999
3.0	0.1	1.062	0.988	1.049	0.883	0.938
	0.3	1.955	0.561	1.098	0.505	0.987
	0.5	3.562	0.305	1.087	0.280	0.997
	0.7	7.312	0.145	1.062	0.137	0.999
	0.9	26.062	0.039	1.025	0.038	1.000

more efficient than $\underline{d}(\hat{\rho})$ or $\underline{\tilde{d}}$. The region of ρ within which $\underline{d}$ is more efficient than $\underline{d}(\hat{\rho})$ or $\underline{\tilde{d}}$ depends upon the magnitude of λ_k. This superiority of $\underline{d}$ vanishes as either T or n increases. For example, when $\rho \leq 0.1$ and $T = 3$, the estimator $\underline{d}(\hat{\rho})$ is more efficient than $\underline{d}$ only for values of n sufficiently larger than 19. When $\rho < 0.1$ and $n = 7$, T should be around 15 for $\underline{d}(\hat{\rho})$ to be more efficient than $\underline{d}$. We should set $\underline{\alpha} = (1,0,1,1)'$ in (2.4.11) when T and ρ are small and n is not very large.

When n is small and the true value of ρ is large (i.e., the value of ρ is around 0.8), $\underline{\tilde{d}}$ is more efficient than $\underline{d}(\hat{\rho})$ or $\underline{d}$ for some values of λ_k. This is true whether T is small or large. The region of ρ within which $\underline{\tilde{d}}$ is superior to $\underline{d}(\hat{\rho})$ or $\underline{d}$ is wider for larger values of λ_k when the value of T is fixed or for larger values of T when the value of λ_k is fixed. The superiority of $\underline{\tilde{d}}$ over $\underline{d}(\hat{\rho})$ for a large value of ρ vanishes as n increases. These results indicate that when n is small and T or $\rho (< 1)$ is large, the precision of the estimator $\hat{\sigma}^2$ will be low and the estimator $\underline{d}(\hat{\rho})$ based on $\hat{\sigma}^2$ will be less efficient than $\underline{\tilde{d}}$. We should set $\underline{\alpha} = (0,\alpha_1,\alpha_2,\alpha_3)'$ in (2.4.11) when n-K is small and the true value of $\rho (< 1)$ or T is large.

In all other cases $\underline{d}(\hat{\rho})$ is more efficient than $\underline{d}$ or $\underline{\tilde{d}}$ and we can set $\underline{\alpha} = (1,1,0,1)'$ in (2.4.11).

The practical conclusion that emerges from the study of values in Table 2.1 is to use $\underline{d}$ in small samples unless the data strongly indicate a large ρ value and then to use $\underline{\tilde{d}}$ or if it is worth the effort $\underline{d}(\hat{\rho})$.

If the individual effect μ_i in (2.4.2) is not random but fixed and different for different units, then we can easily show that $\underline{\tilde{d}}$ is the BLUE of $\underline{\delta}$.

The perceptive reader might have already noticed that the estimator $\underline{\bar{d}}$ in (2.4.12) cannot surpass both the estimators $\underline{d}$ and $\underline{\tilde{d}}$ in efficiency for the same set of values of parameters and sample sizes and therefore can be ignored. The variance-covariance matrix of $\underline{\bar{d}}$ is $\bar{\sigma}^2 PP'$. The estimator $\underline{\bar{d}}$ is more efficient than $\underline{\tilde{d}}$ if $c\lambda_m < 1$ where λ_m is the largest diagonal element of $\Lambda_{K'}$. Let $(1+\lambda_k c^{-1})(1+\lambda_k)^{-2}$ be the largest diagonal element of $D_2 c^{-1}$. The estimator $\underline{d}$ is more efficient than

$\bar{d}$ if $(1+\lambda_k c^{-1})(1+\lambda_k)^{-2} < 1$. If $\lambda_m < 1$ and c is close to 1 such that $c\lambda_m < 1$, then $\bar{d}$ is more efficient than $\tilde{d}$. But for these values of c and λ_k, $\underline{d}$ is more efficient than $\bar{d}$. Thus, $\bar{d}$ cannot beat both $\underline{d}$ and $\tilde{d}$ at the same time.

(d) Asmyptotic Properties of $\underline{d}(\hat{\rho}, \underline{\alpha})$

In discussing the asymptotic properties of $\underline{\alpha}$-I class, it is first necessary to clarify the concept of "large samples" when the sample has two dimensions. Wallace and Hussain (1969) proved the asymptotic properties of the estimators of parameters in an error components model by allowing $n\to\infty$, $T\to\infty$. Our assumptions are of a similar nature.

In this subsection we make use of both Assumption 2.4.2 and the following assumption.[13]

Assumption 2.4.3:

(1) $\lim\limits_{\substack{n\to\infty \\ T\to\infty}} \frac{1}{n}\sum\limits_{i=1}^{n} \frac{X_i'\iota_T}{T}$ is finite and $\lim\limits_{n\to\infty}\left\{\frac{1}{n}\sum\limits_{i=1}^{n}\left(\lim\limits_{T\to\infty}\frac{X_i'\iota_T\iota_T'X_i}{T^2}\right)\right\}$

is a finite, positive definite matrix;

(2) $\operatorname*{plim}\limits_{n\to\infty}\left\{\frac{1}{n}\sum\limits_{i=1}^{n}\left(\operatorname*{plim}\limits_{T\to\infty}\frac{X_i'\iota_T\iota_T'u_i}{T^2}\right)\right\} = 0$;

(3) $\lim\limits_{n\to\infty}\left\{\frac{1}{n}\sum\limits_{i=1}^{n}\left(\lim\limits_{T\to\infty}\frac{X_i'X_i}{T}\right)\right\}$ is finite and positive definite; and

(4) $\operatorname*{plim}\limits_{n\to\infty}\left\{\frac{1}{n}\sum\limits_{i=1}^{n}\left(\operatorname*{plim}\limits_{T\to\infty}\frac{X_i'u_i}{T}\right)\right\} = 0$.

Notice that the specifications (1) and (3) in Assumption 2.4.3 imply that $\lim\limits_{\substack{n\to\infty \\ T\to\infty}} \bar{Z}'N\bar{Z}/nT$ and $\lim\limits_{\substack{n\to\infty \\ T\to\infty}} \tilde{Z}'\tilde{Z}/nT$ are finite, positive matrices. Under Assumption 2.4.3

---

[13] The results that follow could be reproduced without the normality assumption. It would then be necessary to make a large number of lesser assumptions about the moments of the error distribution, cf. Wallace and Hussain (1969).

we have

$$\underset{\substack{n\to\infty\\T\to\infty}}{\text{plim}}\ \hat{\bar{\sigma}}^2/T = \sigma^2\rho \quad , \quad \underset{\substack{n\to\infty\\T\to\infty}}{\text{plim}}\ \hat{\sigma}_\nu^2 = \sigma_\nu^2 \quad , \quad \text{and} \quad \underset{\substack{n\to\infty\\T\to\infty}}{\text{plim}}\ \hat{\rho} = \rho \quad . \tag{2.4.28}$$

We can write

$$\sqrt{nT}\ [\underline{d}(\rho)-\underline{\delta}] = \left[\frac{\bar{Z}'N_1\bar{Z}/nT^2}{\bar{\sigma}^2/T} + \frac{\tilde{Z}'\tilde{Z}}{nT\sigma_\nu^2}\right]^{-1} \left[\frac{\bar{Z}'N_1\bar{\underline{u}}/T\sqrt{nT}}{\bar{\sigma}^2/T} + \frac{\tilde{Z}'\tilde{\underline{u}}}{\sqrt{nT}\ \sigma_\nu^2}\right] \quad . \tag{2.4.29}$$

Since $\lim_{\substack{n\to\infty\\T\to\infty}} \bar{Z}'N_1\bar{Z}/nT^2 = 0$, $\lim_{T\to\infty} \bar{\sigma}^2/T = \sigma_\mu^2$, and $\dfrac{\bar{Z}'N_1\bar{\underline{u}}/T\sqrt{nT}}{\bar{\sigma}^2/T}$ converges in probability to zero, the limiting distribution of $\underline{d}(\rho)$ is the same as that of $\tilde{\underline{d}}$. Similarly,

$$\sqrt{nT}[\underline{d}(\hat{\rho},\underline{\alpha})-\underline{\delta}] = \left[\frac{\alpha_0\bar{Z}'N_1\bar{Z}/nT^2}{(\alpha_1\hat{\bar{\sigma}}^2 + \alpha_2\hat{\sigma}_\nu^2)/T} + \frac{\alpha_3\tilde{Z}'\tilde{Z}}{nT\ \hat{\sigma}_\nu^2}\right]^{-1}$$

$$\cdot \left[\frac{\alpha_0 Z'N_1\bar{\underline{u}}/T\sqrt{nT}}{(\alpha_1\hat{\bar{\sigma}}^2 + \alpha_2\hat{\sigma}_\nu^2)/T} + \frac{\alpha_3\tilde{Z}'\tilde{\underline{u}}}{\sqrt{nT}\ \hat{\sigma}_\nu^2}\right] \quad . \tag{2.4.30}$$

Noting that $\lim_{\substack{n\to\infty\\T\to\infty}} \bar{Z}'N_1\bar{Z}/nT^2 = 0$, we can easily show that when

$$\alpha_1 \neq 0, \quad \underset{\substack{n\to\infty\\T\to\infty}}{\text{plim}}\ (\alpha_1\hat{\bar{\sigma}}^2 + \alpha_2\hat{\sigma}_\nu^2)/T = \alpha_1\sigma_\mu^2 \ , \quad \text{and} \quad \underset{\substack{n\to\infty\\T\to\infty}}{\text{plim}} \left[\frac{\alpha_0\bar{Z}'N_1\bar{Z}/nT^2}{(\alpha_1\hat{\bar{\sigma}}^2 + \alpha_2\hat{\sigma}_\nu^2)/T} + \frac{\alpha_3\tilde{Z}'\tilde{Z}}{nT\ \hat{\sigma}_\nu^2}\right] \tag{2.4.31}$$

is equal to $\underset{\substack{n\to\infty\\T\to\infty}}{\text{plim}} \dfrac{\alpha_3\tilde{Z}'\tilde{Z}}{nT\ \hat{\sigma}_\nu^2}$ which is finite and positive definite provided $\alpha_3 \neq 0$.

Therefore, the asymptotic distribution of the sampling error $\sqrt{nT}\ [\underline{d}(\hat{\rho},\underline{\alpha})-\underline{\delta}]$ is the same as that of the inverse of limit matrix (2.4.31) postmultiplied by

$$\left[\frac{\alpha_0\bar{Z}'N_1\bar{\underline{u}}/T\sqrt{nT}}{(\alpha_1\hat{\bar{\sigma}}^2 + \alpha_2\hat{\sigma}_\nu^2)/T} + \frac{\alpha_3\tilde{Z}'\tilde{\underline{u}}}{\sqrt{nT}\ \hat{\sigma}_\nu^2}\right] \quad . \tag{2.4.32}$$

The first term in (2.4.32) converges in probability to zero and the asymptotic distribution of $\underline{d}(\hat{\rho},\underline{\alpha})$ is the same as that of $\tilde{\underline{d}}$. From (2.4.29) it follows that when $\alpha_1 \neq 0$ and $\alpha_3 \neq 0$, both the estimators $\underline{d}(\hat{\rho},\underline{\alpha})$ and $\tilde{\underline{d}}$ have the same limiting distribution as $\underline{d}(\rho)$ which is the BLUE of $\underline{\delta}$. Since the OLS estimator $\underline{d}$ is obtained from $\underline{d}(\hat{\rho},\underline{\alpha})$ by setting $\alpha_0 = 1$, $\alpha_1 = 0$, $\alpha_2 = 1$ and $\alpha_3 = 1$, it cannot be asymptotically as efficient as $\tilde{\underline{d}}$. The variance-covariance matrix of $\sqrt{nT}(\underline{d}-\underline{\delta})$ is

$$\left[\frac{\bar{Z}'N_1\bar{Z}}{nT} + \frac{\tilde{Z}'\tilde{Z}}{nT}\right]^{-1} \left[\frac{\bar{\sigma}^{-2}\bar{Z}'N_1\bar{Z}}{nT} + \frac{\sigma_\nu^2\tilde{Z}'\tilde{Z}}{nT}\right] \left[\frac{\bar{Z}'N_1\bar{Z}}{nT} + \frac{\tilde{Z}'\tilde{Z}}{nT}\right]^{-1} \quad . \qquad (2.4.33)$$

Since $\bar{\sigma}^{-2}\to\infty$ as $T\to\infty$, the asymptotic variances of the elements of $\underline{d}$ are unbounded. In the same way we can also show that the elements of $\sqrt{nT}\,[\underline{d}(\hat{\rho},\underline{\alpha})-\underline{\delta}]$ with $\underline{\alpha} = (\alpha_0 \neq 0,0,1,\alpha_3 \neq 0)'$ or with $\underline{\alpha} = (\alpha_0 \neq 0,\alpha_1,\alpha_2,0)'$ have unbounded asymptotic variances.[14] This shows that $\bar{\underline{d}}$ in (2.4.12) is an asymptotically inefficient estimator. Any $\underline{\alpha}$-estimator with $\underline{\alpha} = (\alpha_0,\alpha_1 \neq 0, \alpha_2, \alpha_3 \neq 0)'$ has an asymptotic distribution which is the same as that of $\underline{d}(\rho)$. The point to be made in this subsection is that the set of asymptotically efficient estimators (i.e., the estimators whose limiting distributions are the same as that of $\underline{d}(\rho)$) is infinite. It is not possible to choose an estimator of $\underline{\delta}$ on the basis of asymptotic efficiency. The small sample properties of different estimators of $\underline{\delta}$, which we established in subsection 2.4(c), provide a basis for choosing a feasible efficient estimator of $\underline{\delta}$.

Notice that the specifications (1) and (3) in Assumption 2.4.3 imply that $\lim_{\substack{n\to\infty \\ T\to\infty}} \lambda_k$ is finite and positive. Recognizing that $\operatorname*{plim}_{\substack{n\to\infty \\ T\to\infty}} f/c = 1$ we can easily show that $\operatorname*{plim}_{\substack{n\to\infty \\ T\to\infty}} e_{1m} = 0$, $\operatorname*{plim}_{\substack{n\to\infty \\ T\to\infty}} e_{2m} = 1$, and $\lim_{\substack{n\to\infty \\ T\to\infty}} e_{3m} = \infty$.

---

[14]Notice, however, that $\sqrt{n}[\underline{d}(\hat{\rho},\underline{\alpha})-\underline{\delta}]$ for $\underline{\alpha} = (\alpha_0 \neq 0, \alpha_1, \alpha_2, 0)'$ has a finite asymptotic variance-covariance matrix.

(e) <u>Alternative Estimators of Coefficients and Relationship to Previous Results</u>

The $\underline{\alpha}$-I class of estimators in (2.4.11) are based on the variance estimators $\hat{\bar{\sigma}}^2$ and $\hat{\sigma}_\nu^2$ which are obtained by using two different estimators of the same slope coefficients. We ignored the restriction that the slope coefficients in (2.4.6a) and (2.4.6b) are the same while deriving the estimators of $\bar{\sigma}^2$ and σ_ν^2. We can incorporate this restriction by following Rao's (1970) principle of Minimum Norm Quadratic Unbiased Estimation (MINQUE). According to this principle estimates of $\bar{\sigma}^2$ and σ_ν^2 are obtained as solutions to the equations given by

$$a_{11}\bar{s}^2 + a_{12}s_\nu^2 = \underline{w}_1'\underline{w}_1$$

$$a_{21}\bar{s}^2 + a_{22}s_\nu^2 = \underline{w}_2'\underline{w}_2 \quad , \qquad (2.4.34)$$

where

$$a_{11} = \mathrm{tr}[I_{n'} - C_2N_1\bar{Z}(\bar{Z}'N_1\bar{Z}+\tilde{Z}'\tilde{Z})^{-1}\bar{Z}'N_1C_2']^2$$

$$a_{12} = a_{21} = \mathrm{tr}C_2N_1\bar{Z}(\bar{Z}'N_1\bar{Z}+\tilde{Z}'\tilde{Z})^{-1}\tilde{Z}'\tilde{Z}(\bar{Z}'N_1\bar{Z}+\tilde{Z}'\tilde{Z})^{-1}\bar{Z}'N_1C_2'$$

$$a_{22} = \mathrm{tr}[I_{nT'} - \tilde{Z}(\bar{Z}'N_1\bar{Z}+\tilde{Z}'\tilde{Z})^{-1}\tilde{Z}']^2$$

$$C_2N_1C_2' = I_{n'} \text{ such that } N_1 = N_1C_2'C_2N_1 \quad ,$$

$$\underline{w}_1 = C_2N_1\bar{\underline{y}} - C_2N_1\bar{Z}(\bar{Z}'N_1\bar{Z}+\tilde{Z}'\tilde{Z})^{-1}(\bar{Z}'N_1\bar{\underline{y}} + \tilde{Z}'\tilde{\underline{y}}) \text{ , and}$$

$$\underline{w}_2 = \tilde{\underline{y}} - \tilde{Z}(\bar{Z}'N_1\bar{Z}+\tilde{Z}'\tilde{Z})^{-1}(\bar{Z}'N_1\bar{\underline{y}} + \tilde{Z}'\tilde{\underline{y}}) \quad .$$

Notice that C_2 is not unique but the values of a_{11}, a_{12}, a_{21}, and a_{22} are not affected by the arbitrariness of C_2. Suppose that $A = (a_{ij})$ is nonsingular. Then the solutions to (2.4.34) are

$$\begin{bmatrix} \bar{s}^2 \\ s_\nu^2 \end{bmatrix} = A^{-1} \begin{bmatrix} \underline{w}_1'\underline{w}_1 \\ \underline{w}_2'\underline{w}_2 \end{bmatrix} \quad . \qquad (2.4.35)$$

By substituting these estimates of $\bar{\sigma}^2$ and σ_ν^2 in the Aitken estimator $\underline{d}(\rho)$, we obtain an alternative estimator of $\underline{\delta}$. Such an estimator may still be unbiased and more efficient than any member of $\underline{\alpha}$-I class for some values of the parameters in small samples; however, a detailed investigation is necessary. A theoretical study of this problem is difficult. Monte Carlo studies may throw some light. In large samples we are not too hopeful for an Aitken procedure based on $\bar{s}^2$ and s_ν^2 because $\underline{w}_1$ and $\underline{w}_2$ are essentially OLS residuals and the OLS estimator of $\underline{\delta}$ is asymptotically inefficient. The efficiency of coefficient estimators affect the efficiency of variance estimators and vice-versa.

We can also follow an iterative ML method developed by Nerlove (1965, pp. 157-187) and applied by Balestra and Nerlove (1966) in their study on the demand for natural gas in the U.S. In small samples restrictions on coefficients and variances may be important and if they are utilized in any estimation procedure, an estimator that comes out of such a procedure may be more efficient than any other estimator obtained by ignoring those restrictions. In large samples prior information will be dominated by sample information and the prior restrictions on the parameters, when incorporated, cannot improve the asymptotic efficiency of an estimator of $\underline{\beta}$. Consequently, it may not be possible to choose between an iterative ML estimator and a member of $\underline{\alpha}$-I class on the basis of asymptotic efficiency.

A Bayesian approach was followed by Tiao and Draper (1968) in analyzing a linear model of the type (2.4.1) and (2.4.2). Combining the likelihood function of parameters in (2.4.1) with noninformative diffuse priors on $\underline{\beta}$, $\bar{\sigma}^2$ and σ_ν^2 via Bayes' theorem, Tiao and Draper obtained the conditional posterior density of $\underline{\delta}$ given ρ, which was centered at the sampling theory quantity $\underline{d}(\rho)$. This conditional posterior density provided a good approximation to the marginal posterior density of $\underline{\delta}$ under certain conditions. They also noted that the restriction $\bar{\sigma}^2 \geq \sigma_\nu^2$ had the effect of pulling the marginal posterior density of $\underline{\delta}$ towards a distribution centered at $\underline{\tilde{d}}$. In our sampling theory approach by a prudent choice of constants $\underline{\alpha}$ which would increase the weight given to $\underline{\tilde{d}}$ and decrease the weight given to $\bar{\underline{d}}$ in (2.4.12), we may be able to make $\underline{d}(\hat{\rho},\underline{\alpha})$ surpass $\underline{d}(\hat{\rho})$ in efficiency for some finite values of n,T,K,ρ and λ_k. Following a numerical integration procedure we evaluated the variance of an element of $\underline{\alpha}$-estimator with $\underline{\alpha} = (1,1,0.4,1)'$ but this variance was only slightly smaller than the variance of the corresponding element of $\underline{d}(\hat{\rho})$.

Nerlove (1965, pp. 157-187) developed a two-round procedure to estimate the parameters in a model of the type (2.4.1) and (2.4.2) with stochastic regressors. This procedure which requires T to be greater than K, is already described in subsection 1.2(a). The variance estimators (1.2.5d) and (1.2.5e) are different from the estimators of σ_μ^2 and σ^2, implied by (2.4.8) and (2.4.9) and are justified only on asymptotic grounds. Nerlove (1967) designed a Monte Carlo experiment to study the small sample properties of alternative estimators. He considered a simple dynamic equation, $y_t = \gamma_1 y_{t-1} + u_t$ without exogenous variables. His experimental results confirmed the superiority of the two-round procedure of estimating γ_1 over the OLS in all cases considered by him and also over the ML procedure in some cases. If the lagged values of the dependent variable appear as independent variables in (2.4.1), then the assumption of full independence between regressors and disturbances is untenable. The regressors will be correlated with μ_i and with all the past values of ν_{it} but not with the current and future values of ν_{it} since the elements of $\underline{\nu}_i$ are assumed to be serially independent, cf. Goldberger (1964, p. 272). Therefore, the assumption that $\operatorname{plim}_{\substack{T\to\infty \\ n\to\infty}} \sum_{i=1}^{n} X_i'\underline{\nu}_i/Tn = 0$ is tenable. Replacing "lim" by "plim" in the specifications (1) and (3) of Assumption 2.4.3, we can easily show that Assumption 2.4.3 is valid even when the lagged values of the dependent variable appear on the r.h.s. of (2.4.1), cf. Nerlove (1965, pp. 180-2). Consequently, our estimators $\hat{\rho}$ and $\underline{b}(\hat{\rho})$ are consistent. In this case also the estimators $\tilde{\underline{d}}$ and $\underline{d}(\hat{\rho})$ will have the same limiting distribution. Thus, the asymptotic results established in subsection 2.4(d) are valid even when the lagged values of the dependent variable occur as explanatory variables provided the elements of $\underline{\nu}_i$ are serially independent. Needless to say that our small sample results are not valid in this case. In large samples we can adopt the estimator

$\underline{d}(\hat{\rho})$ or $\tilde{\underline{d}}$ to estimate the parameters in the dynamic equation considered by Balestra and Nerlove whether $T < K$ or $T > K$ provided $n > K$ and $T > 1$. The estimators (2.4.35), which are based on OLS residuals, are inconsistent when the lagged values of the dependent variable appear as explanatory variables and they should not be used in this case, cf. Nerlove (1965, p. 164).

2.5 Estimation When the Variance-Covariance Matrix of Disturbances is Singular

If we are given *a priori* that $\sigma_\nu^2 = 0$ or $\rho = 1$, Ω is singular and the Aitken estimator of $\underline{\beta}$ does not exist. The prior information that $\rho = 1$ and $\sigma^2 \neq 0$ implies that $E\tilde{\underline{u}} = 0$, $E\tilde{\underline{u}}\tilde{\underline{u}}' = 0$ and the equation (2.4.6b) is an exact relationship. Furthermore, in the equation (2.4.6a) $E\bar{\underline{u}}\bar{\underline{u}}' = T\sigma^2 I$. An efficient procedure of estimating $\underline{\beta}$ in this case is to minimize $\bar{\underline{u}}'\bar{\underline{u}}$ subject to the condition $\tilde{\underline{y}} = \tilde{Z}'\underline{\delta}$. In order to make the *a priori* restrictions consistent we premultiply both sides by $\tilde{Z}'$ to obtain $\tilde{Z}'\tilde{\underline{y}} = \tilde{Z}'\tilde{Z}\underline{\delta}$. For the entire coefficient vector $\underline{\beta}$ we write the restrictions as

$$\tilde{\underline{r}} = \tilde{R}\underline{\beta} \quad , \tag{2.5.1}$$

where $\tilde{\underline{r}} = \tilde{Z}'\tilde{\underline{y}}$ and $\tilde{R} = [0, \tilde{Z}'\tilde{Z}]$ is a (K-1) XK matrix of known constants. We assume that the matrix $\tilde{R}$ has full row rank.

Following the procedure outlined in Goldberger (1964, pp. 256-7) we have

$$\underline{b}^* = (\bar{X}'\bar{X})^{-1}\bar{X}'\bar{\underline{y}} + (\bar{X}'\bar{X})^{-1}\tilde{R}'[\tilde{R}(\bar{X}'\bar{X})^{-1}\tilde{R}']^{-1}[\tilde{\underline{r}} - \tilde{R}(\bar{X}'\bar{X})^{-1}\bar{X}'\bar{\underline{y}}] \tag{2.5.2}$$

as an estimator of $\underline{\beta}$. The properties of $\underline{b}^*$ are well known, cf. Goldberger (1964, pp. 256-7).

2.6 Estimation When the Remaining Effects are Heteroskedastic

(a) Aitken's Generalized Least Squares Using the Estimated Variances of Disturbances

Consider the basic equation (2.4.1) and the error specification (2.4.2). Assumption 2.4.1 is modified as follows.

<u>Assumption 2.6.1</u>:

(1) The sample sizes n and T are such that $n > K$ and $T > K$.

(2) Same as (2) in Assumption 2.4.1.

(3) Same as (3) in Assumption 2.4.1.

(4) The $\underline{\nu}_i$'s are independently distributed with the same mean vector zero. The variance-covariance matrix of $\underline{\nu}_i$ is $\sigma_{ii}I_T$.

(5) Same as (5) in Assumption 2.4.1.

This assumption is slightly more general than Assumption 2.4.1 because it allows the common variance of the elements of $\underline{\nu}_i$ to vary across units. Using Assumption 2.6.1 we can show that

$$E\underline{u}_i\underline{u}_i' = \sigma_\mu^2\underline{\iota}_T\underline{\iota}_T' + \sigma_{ii}I_T \quad , \quad E\bar{u}_i^2 = T\sigma_\mu^2 + \sigma_{ii} \quad , \quad E\tilde{\underline{u}}_i\tilde{\underline{u}}_i' = \sigma_{ii}I_T \quad ,$$

and $E\underline{u}_i\underline{u}_j' = 0$ for $i \neq j$. Since $T > K$, an unbiased estimator of σ_{ii} is given by

$$\tilde{\sigma}_{ii} = \tilde{\underline{y}}_i'\tilde{M}_i\tilde{\underline{y}}_i/(T-K) \qquad (i = 1,2,\ldots,n) \quad , \tag{2.6.1}$$

where $\tilde{M}_i = I_{T'} - \tilde{Z}_i(\tilde{Z}_i'\tilde{Z}_i)^{-1}\tilde{Z}_i'$, $\tilde{\underline{y}}_i = C_1\underline{y}_i$, $\tilde{Z}_i = C_1Z_i$, $X_i = [\underline{\iota}_T, Z_i]$, $0_T = [\underline{\iota}_T/\sqrt{T}, C_1']'$ is an orthogonal matrix, and Z_i is as explained in (2.3.1). Taking the expectation on both sides of (2.4.8) under Assumption 2.6.1 we have

$$E\hat{\bar{\sigma}}^2 = E\bar{\underline{u}}'\bar{M}\bar{\underline{u}}/(n-K) = (n-K)^{-1}[E\bar{\underline{u}}'\bar{\underline{u}} - E\bar{\underline{u}}'\bar{X}(\bar{X}'\bar{X})^{-1}\bar{X}'\bar{\underline{u}}] \tag{2.6.2}$$

and

$$E\bar{\underline{u}}\bar{\underline{u}}' = \mathrm{diag}(T\sigma_\mu^2 + \sigma_{11},\ldots,T\sigma_\mu^2 + \sigma_{nn}) \quad . \tag{2.6.3}$$

Substituting (2.6.3) in (2.6.2) we have

$$E\hat{\bar{\sigma}}^2 = (n-K)^{-1}\Big[nT\sigma_\mu^2 + \sum_{i=1}^n \sigma_{ii} - TK\,\sigma_\mu^2 - \mathrm{tr}(\bar{X}'\bar{X})^{-1} \cdot \Big(\sum_{i=1}^n \sigma_{ii}T^{-1}X_i'\underline{\iota}_T\underline{\iota}_T'X_i\Big)\Big] \quad . \tag{2.6.4}$$

Consequently, an unbiased estimator of σ_μ^2 is given by

$$\hat{\sigma}_\mu^2 = T^{-1}\hat{\hat{\sigma}}^2 - [T(n-K)]^{-1}\Big[\sum_{i=1}^n \tilde{\sigma}_{ii} - \mathrm{tr}(\bar{X}'\bar{X})^{-1}$$

$$\cdot\Big(\sum_{i=1}^n \tilde{\sigma}_{ii}T^{-1}X_i'\,\underline{\iota}_T\underline{\iota}_T'X_i\Big)\Big] \quad . \tag{2.6.5}$$

Again the structure of $\hat{\sigma}_\mu^2$ is such that it can yield negative estimates in some applications. We can compare the structure of $\hat{\sigma}_\mu^2$ in (2.6.5) with that of the numerator of $\hat{\rho}$ in (2.4.10). If the specification (4) in Assumption 2.4.1 is in error and the specification (4) in Assumption 2.6.1 is appropriate to the situation under study, adoption of the estimator $\hat{\sigma}_\mu^2$ may result in a positive estimate while the adoption of $\hat{\rho}$ may result in a negative estimate. Considerations of obtaining plausible estimates for variances provide a guide to the selection of appropriate set of assumptions.

An Aitken estimator of $\underline{\delta}$ based on estimated variances of disturbances is given by

$$\underline{d}(\hat{\underline{\omega}}) = \Big[\sum_{i=1}^n \frac{\bar{\underline{z}}_i^*\bar{\underline{z}}_i^{*\prime}}{(T\hat{\sigma}_\mu^2 + \tilde{\sigma}_{ii})} + \sum_{i=1}^n \frac{\tilde{Z}_i\tilde{Z}_i'}{\tilde{\sigma}_{ii}}\Big]^{-1}\Big[\sum_{i=1}^n \frac{\bar{\underline{z}}_i^*\bar{y}_i^*}{(T\hat{\sigma}_\mu^2 + \tilde{\sigma}_{ii})} + \sum_{i=1}^n \frac{\tilde{Z}_i'\tilde{\underline{y}}_i}{\tilde{\sigma}_{ii}}\Big] \quad , \tag{2.6.6}$$

where $\bar{\underline{z}}_i^{*\prime} = \underline{\iota}_T'(Z_i - n^{-1}\sum_{i=1}^n Z_i)/\sqrt{T}$, $\bar{y}_i^* = \underline{\iota}_T'(\underline{y}_i - n^{-1}\sum_{i=1}^n \underline{y}_i)/\sqrt{T}$, and $\underline{\omega}$ is a $(n+1)\mathrm{x}1$ vector containing σ_μ^2 and $\sigma_{ii}(i = 1,2,\ldots,n)$ as its elements. An estimator of β_0 is given by

$$b_0(\hat{\underline{\omega}}) = (n\sqrt{T})^{-1}\sum_{i=1}^n [\bar{y}_i - \bar{\underline{z}}_i'\underline{d}(\hat{\underline{\omega}})] \quad . \tag{2.6.7}$$

Let $\underline{b}(\hat{\underline{\omega}}) = [b_0(\hat{\underline{\omega}})\ ,\ \underline{d}'(\hat{\underline{\omega}})]'$. Using Assumptions 2.4.2 and 2.6.1 and the results $C_1\underline{\iota}_T = 0$, $\tilde{M}_i\tilde{Z}_i = 0$ and $\bar{M}\bar{X} = 0$, we can show that $\bar{u}_i$ is independent of $\tilde{\underline{u}}_j$ for every i&j, $\tilde{Z}_i'\tilde{\underline{u}}_i$ is independent of $\tilde{\sigma}_{ii}$ for every i, and $\bar{X}'\bar{\underline{u}}$ is independent of $\hat{\hat{\sigma}}^2$. Since $\bar{u}_i^* = \bar{u}_i - n^{-1}\sum_{i=1}^n \bar{u}_i$, $\bar{u}_i = \underline{\iota}_T'\underline{u}_i/\sqrt{T}$, and $\tilde{\underline{u}}_i$ is independent of $\bar{u}_i$, the above results imply that $\bar{\underline{z}}_i^*\bar{u}_i^*$ is independent of $\tilde{\sigma}_{ii}$. $\bar{X}'\bar{M} = 0$ implies $\underline{\iota}_n'\bar{M} = 0$. This means that $n^{-1}\underline{\iota}_n'\bar{\underline{u}}$ is independent of $\hat{\hat{\sigma}}^2$. Unfortunately, $\bar{\underline{z}}_i^*\bar{u}_i$ is <u>not</u> independent of those terms on the r.h.s. of (2.4.8) involving $\bar{u}_i$. This creates difficulties in deriving the exact moments of $\underline{d}(\hat{\underline{\omega}})$. The distribution of $(T-K)\tilde{\sigma}_{ii}/\sigma_{ii}$ is straightforwardly derived as χ^2 with T-K d·f. The derivation of the exact distribution of $\hat{\sigma}_{\mu}^2$ is complicated.

In a compact form we can write

$$\underline{d}(\hat{\underline{\omega}}) = (W_1' \hat{\bar{\Sigma}}_d^{-1} W_1)^{-1} W_1' \hat{\bar{\Sigma}}_d^{-1} \underline{r}_1 \quad , \tag{2.6.8}$$

where $W_1 \equiv [\bar{Z}'N_1, \tilde{Z}']'$ is a nTxK' matrix, $K' = K-1$, $\hat{\bar{\Sigma}}_d = \text{diag}[T\hat{\sigma}_{\mu}^2 + \tilde{\sigma}_{11}, \ldots, T\hat{\sigma}_{\mu}^2 + \tilde{\sigma}_{nn}, \tilde{\sigma}_{11} I_{T'}, \ldots, \tilde{\sigma}_{nn} I_{T'}]$ is a nT x nT diagonal matrix and $\underline{r}_1 \equiv [\bar{\underline{y}}'N_1, \tilde{\underline{y}}']'$. Let

$$\underline{r}_1 = W_1 \underline{\delta} + \underline{v}_1 \quad , \tag{2.6.9}$$

where $\underline{v}_1 = [\bar{\underline{u}}'N_1, \tilde{\underline{u}}']'$. Substituting (2.6.9) into (2.6.8) we obtain

$$\underline{d}(\hat{\underline{\omega}}) - \underline{\delta} = (W_1' \hat{\bar{\Sigma}}_d^{-1} W_1)^{-1} W_1' \hat{\bar{\Sigma}}_d^{-1} \underline{v}_1 \quad . \tag{2.6.10}$$

Now we can follow the argument in Kakwani (1967) to show that $\underline{d}(\hat{\underline{\omega}})$ is an unbiased estimator of $\underline{\delta}$. Under Assumption 2.4.2, $\underline{v}_1$ follows a nT-dimensional normal distribution with mean vector zero and variance-covariance matrix $\bar{\Sigma}_d = \text{diag}[T\sigma_{\mu}^2 + \sigma_{11}, \ldots, T\sigma_{\mu}^2 + \sigma_{nn}, \sigma_{11} I_{T'}, \ldots, \sigma_{nn} I_{T'}]$. This is a symmetric distribution. We note that the matrix $\hat{\bar{\Sigma}}_d$ is an even function of $\underline{v}_1$ in the sense that it is invariant to a change in the sign of all elements of $\underline{v}_1$. For the same reason the matrix $(W_1' \hat{\bar{\Sigma}}_d^{-1} W_1)^{-1} W_1' \hat{\bar{\Sigma}}_d^{-1}$ is an even function of $\underline{v}_1$. Substitution of $-\underline{v}_1$ in (2.6.10) merely changes the sign of $\underline{d}(\hat{\underline{\omega}}) - \underline{\delta}$. This means that $\underline{d}(\hat{\underline{\omega}}) - \underline{\delta}$ has the same probability density function as $\underline{\delta} - \underline{d}(\hat{\underline{\omega}})$ because $\underline{v}_1$ and $-\underline{v}_1$ have the same probability density function. Hence, $\underline{d}(\hat{\underline{\omega}})$ is symmetrically distributed about the value $\underline{\delta}$. This proves that $\underline{d}(\hat{\underline{\omega}})$ is unbiased.

This is as far as we can go with regard to the exact finite sample properties of $\underline{d}(\hat{\underline{\omega}})$. Our study of simpler cases in Section 2.4 can provide some hints as regards the small sample properties of $\underline{d}(\hat{\underline{\omega}})$. If the sample sizes n and T are so large as to provide sufficient d.f. to estimate σ_{μ}^2 and the σ_{ii}'s precisely and if the true value of σ_{μ}^2 is not small and if the variances $\sigma_{11}, \ldots, \sigma_{nn}$ are not equal to one another, then the estimator $\underline{d}(\hat{\underline{\omega}})$ is superior to any other unbiased estimator of $\underline{\delta}$ according to the minimum variance criterion. If the sample sizes n and T, and the true value of σ_μ^2 are small, and if the variances $\sigma_{11},\ldots,\sigma_{nn}$ are almost equal to one another, then the OLS estimator (2.4.13) is superior to $\underline{d}(\hat{\underline{\omega}})$. If the variances $\sigma_{11},\ldots,\sigma_{nn}$ differ appreciably from one another and if the sample size n is so small as to make the estimator $\hat{\sigma}_\mu^2$ imprecise and if the sample size T is so large as to provide enough degrees of freedom to estimate each σ_{ii} precisely, then the following estimator is superior to $\underline{d}(\hat{\underline{\omega}})$ and also to (2.4.13).

$$\underline{d}(\tilde{\underline{\sigma}}) = \left[\sum_{j=1}^{n} \frac{\tilde{Z}_j'\tilde{Z}_j}{\tilde{\sigma}_{jj}}\right]^{-1} \left[\sum_{i=1}^{n} \frac{\tilde{Z}_i'\tilde{\underline{y}}_i}{\tilde{\sigma}_{ii}}\right] , \tag{2.6.11}$$

where $\tilde{\underline{\sigma}}$ is a n x 1 vector containing $\tilde{\sigma}_{11},\ldots,\tilde{\sigma}_{nn}$ as its elements.

Remembering that $\tilde{Z}_i'\tilde{\underline{u}}_i$ is independent of $\tilde{\sigma}_{ii}$ for every i we can easily show that $E[\tilde{\underline{d}}(\tilde{\underline{\sigma}})|\tilde{\sigma}_{ii}(i=1,2,\ldots,n)] = \underline{\delta}$ and $E\tilde{\underline{d}}(\tilde{\underline{\sigma}}) = \underline{\delta}$. If n=2 we can also study the exact variances of the elements of $\tilde{\underline{d}}(\tilde{\underline{\sigma}})$ using the analytical and numerical methods considered in Mehta and Swamy (1970).

(b) Asymptotic Properties of $\underline{b}(\hat{\underline{\omega}})$

Under Assumptions 2.4.2, 2.4.3 and 2.6.1 we can easily show that

$$\operatorname*{plim}_{\substack{n\to\infty \\ T\to\infty}} \hat{\sigma}_\mu^2 = \sigma_\mu^2 \quad \text{and} \quad \operatorname*{plim}_{\substack{n\to\infty \\ T\to\infty}} \tilde{\sigma}_{ii} = \sigma_{ii} \quad . \tag{2.6.12}$$

We write

$$\sqrt{nT}[\underline{d}(\hat{\underline{\omega}})-\underline{\delta}] = \left[\sum_{j=1}^{n} \frac{\bar{\underline{z}}_j^*\bar{\underline{z}}_j^{*\prime}}{nT^2} T(T\hat{\sigma}_\mu^2 + \tilde{\sigma}_{jj})^{-1} + \sum_{j=1}^{n} \frac{\tilde{Z}_j'\tilde{Z}_j}{nT\tilde{\sigma}_{jj}}\right]^{-1}$$

$$\cdot \left[\sum_{i=1}^{n} \frac{\bar{\underline{z}}_i^*\bar{\underline{u}}_i^*}{T\sqrt{nT}} T(T\hat{\sigma}_\mu^2 + \tilde{\sigma}_{ii})^{-1} + \sum_{i=1}^{n} \frac{\tilde{Z}_i'\tilde{\underline{u}}_i}{\sqrt{nT}\tilde{\sigma}_{ii}}\right] .$$

Taking the probability limits on both sides of (2.6.13) and recognizing that $\operatorname*{plim}_{\substack{n\to\infty \\ T\to\infty}} \bar{z}_i^* \bar{z}_i^{*\prime}/nT^2 = 0$, we can easily show that

$$\operatorname*{plim}_{\substack{n\to\infty \\ T\to\infty}} |\underline{d}(\hat{\underline{\omega}}) - \underline{\tilde{d}}(\tilde{\sigma})| = 0 \quad . \tag{2.6.14}$$

It follows from a limit theorem in Rao (1965a, p. 101) that the limiting distribution of $\underline{d}(\hat{\underline{\omega}})$ is the same as that of $\underline{d}(\tilde{\underline{\sigma}})$.

If $(\beta_0 + \mu_i)(i=1,2,\ldots,n)$ are fixed parameters, then (2.6.11) is an Aitken estimator of $\underline{\delta}$ based on estimated variances.

2.7 Conclusions

The discussion in this chapter deals with a regression model, wherein the disturbances have a constant variance and pairwise equal correlation. This model is closely related to a regression equation with random intercepts and fixed slopes. We have indicated the conditions under which different estimators of a coefficient are efficient according to the minimum variance criterion. If the variances of disturbances are unknown, we cannot find a single estimator of the coefficient vector which is uniformly better than other estimator according to the minimum variance criterion.

APPENDIX TO CHAPTER II

A. Derivation of the Exact Finite Sample Distribution of the Estimator $\underline{d}(\hat{\rho})$

In order to evaluate the expectation of D(f) in (2.4.18), we first derive the distribution of f. Since $\bar{\underline{u}}$ is normally distributed and $\bar{M}$ is an idempotent matrix, $(n-K)\,\hat{\bar{\sigma}}^2/\bar{\sigma}^2$ is χ^2-distributed with p=(n-K) d.f. Similarly, [n(T-1)-K+1] $\cdot\hat{\sigma}_\upsilon^2/\sigma_\upsilon^2$ is χ^2-distributed with q = [n(T-1)-K+1] d.f. Since $\bar{\underline{u}}$ and $\tilde{\underline{u}}$ are independently distributed, $\hat{\bar{\sigma}}^2$ is independent of $\hat{\sigma}_\upsilon^2$. Consequently, the variable $\hat{\bar{\sigma}}^2\sigma_\upsilon^2/\hat{\sigma}_\upsilon^2\bar{\sigma}^2$ is distributed like F with p, q d.f. Now the question is: Is it correct to take the range of the variable F from 0 to ∞ when we know a priori that $\bar{\sigma}^2 \geq \sigma_\upsilon^2$ and $p < q$? The range of F will be from 1 to ∞ if these conditions imply that $\text{pr}(f \geq c) = 1$. We have $\text{pr}(f \geq c) = \text{pr}[(\hat{\bar{\sigma}}^2 - c\,\hat{\sigma}_\upsilon^2) \geq 0] = \text{pr}(\underline{u}'[p^{-1}Q_1'\bar{M}Q_1 - q^{-1}cJ'\tilde{M}J]\underline{u} \geq 0)$ where $\underline{u} \equiv [\underline{u}_1', \underline{u}_2', \ldots, \underline{u}_n']'$, $Q_1 = (I_n \otimes \underline{\iota}_T'/\sqrt{T})$ is a nXnT matrix and $J \equiv (I_n \otimes C_1)$ is a n(T-1) XnT matrix. If the matrix $(p^{-1}Q_1'\bar{M}Q_1 - q^{-1}cJ'\tilde{M}J)$ is nonnegative definite, then $\text{pr}(f \geq c) = 1$, cf. Box (1954). It is not possible to show that the matrix $p^{-1}Q_1'\bar{M}Q_1 - q^{-1}cJ'\tilde{M}J$ is always nonnegative definite even when $\bar{\sigma}^2 \geq \sigma_\upsilon^2$ and $p < q$. The conditions $\bar{\sigma}^2 \geq \sigma_\upsilon^2$ and $p < q$ only imply that $E\hat{\bar{\sigma}}^2 \geq E\hat{\sigma}_\upsilon^2$ and $\text{var}(\hat{\bar{\sigma}}^2) \geq \text{var}(\hat{\sigma}_\upsilon^2)$. They do not imply that $\text{pr}(f \geq c) = 1$.* Therefore, even when $\bar{\sigma}^2 \geq \sigma_\upsilon^2$, ρ can assume negative values with positive probability. We take the range of F from 0 to ∞. The probability density function (pdf) of $F = \hat{\bar{\sigma}}^2\sigma_\upsilon^2/\hat{\sigma}_\upsilon^2\bar{\sigma}^2$ is

$$p(f) = \frac{(ac)^{-p/2}}{B\left(\frac{p}{2}, \frac{q}{2}\right)} \frac{f^{\frac{p}{2}-1}}{\left(1 + \frac{f}{ac}\right)^{(p+q)/2}} \qquad 0 < f < \infty \quad , \tag{A.1}$$

where $a = q/p$ and $B(.,.)$ is a beta function, cf. Rao (1965a, p. 135). In order to make the variance of $F^{-1} = \hat{\sigma}_\upsilon^2\bar{\sigma}^2/\hat{\bar{\sigma}}^2\sigma_\upsilon^2$ finite we assume that $p > 4$. This imposes the restriction that $n > K + 4$. Now

---

* This difficulty does not arise in Bayesian analysis because in that $\bar{\sigma}^2/\sigma^2$ is a random variable and its range can be taken as 1 to ∞, cf. Tiao and Draper (1968). We also do not face this difficulty in sampling theory approach if $-(T-1)^{-1} < \rho < 1$.

$$E\ \frac{[\alpha_0^2(1+T\xi) + \alpha_3^2\ \lambda_k(\alpha_1 f+\alpha_2)^2]}{[\alpha_0+\alpha_3\lambda_k(\alpha_1 f+\alpha_2)]^2} = \frac{(ac)^{q/2}}{\alpha_1^2\ \lambda_k^2\ \alpha_3^2\ B\left(\frac{p}{2},\frac{q}{2}\right)}$$

$$\cdot\int_0^\infty \frac{[\alpha_0^2(1+T\xi) + \alpha_3^2\ \lambda_k(\alpha_1 f+\alpha_2)^2]}{(\alpha_0\alpha_3^{-1}\lambda_k^{-1}\alpha_1^{-1} + \alpha_2\alpha_1^{-1} + f)^2}\ \frac{f^{(p/2)-1}}{(ac+f)^{(p+q)/2}}\ df. \qquad (A.2)$$

When (p+q)/2 is even the integral (A.2) can be evaluated by resolving the integrand into partial fractions, cf. Gradshteyn and Ryzhik (1965, pp. 56-57). Let (p+q)/2 = 2m. Then the value of the integral on the r.h.s. of (A.2) is

$$\frac{(ac)^{q/2}}{\alpha_1^2\lambda_k^2\alpha_3^2\ B\left(\frac{p}{2},\frac{q}{2}\right)}\left[A_1\ \log\frac{ac\alpha_1\alpha_3\lambda_k}{(\alpha_0+\alpha_2\lambda_k\alpha_3)} + \frac{A_2\lambda_k\alpha_1\alpha_3}{(\alpha_0+\alpha_2\alpha_3\lambda_k)}\right.$$

$$\left. + \frac{B_2}{ac} + \frac{B_3}{2(ac)^2} + \ldots + \frac{B_{2m}}{(2m-1)\ (ac)^{2m-1}}\right], \qquad (A.3)$$

where

$$\psi_1(f) = \frac{[\alpha_0^2(1+T\xi) + \alpha_3^2\ \lambda_k(\alpha_1 f+\alpha_2)^2]f^{(p/2)-1}}{(ac+f)^{2m}},$$

$$\psi_2(f) = \frac{[\alpha_0^2(1+T\xi) + \alpha_3^2\ \lambda_k(\alpha_1 f+\alpha_2)^2]f^{(p/2)-1}}{(\alpha_0\alpha_3^{-1}\lambda_k^{-1}\alpha_1^{-1} + \alpha_2\alpha_1^{-1} + f)^2},$$

$$A_2 = [\psi_1(f)]f = -(\alpha_0\alpha_3^{-1}\lambda_k^{-1}\alpha_1^{-1} + \alpha_2\alpha_1^{-1}),$$

$$A_1 = \left[\frac{d\psi_1(f)}{df}\right]_{f\ =\ -(\alpha_0\alpha_3^{-1}\alpha_1^{-1}\lambda_k^{-1} + \alpha_2\alpha_1^{-1})},$$

$$B_{2m} = [\psi_2(f)]_{f\ =\ -ac} \quad \text{and} \quad B_{2m-i} = \frac{1}{i!}\left[\frac{d^i\psi_2(f)}{dF^i}\right]_{f\ =\ -ac}$$

$$(i=1,2,\ldots,2m-2).$$

When (p+q)/2 is odd, (A.3) involves an infinite series; consequently, we do not consider it here. This shows that ED(f) in (2.4.18) is finite.

The conditional pdf of $\underline{d}(\hat{\rho})$ given f is normal because $[\underline{d}(\hat{\rho})-\underline{\delta}]$, for fixed f, is linear in the independent normal components of the vectors $\underset{\sim}{u}$ and $\underline{u}$.

$$p[\underline{d}(\hat{\rho})|f] = (2\pi)^{-\frac{K-1}{2}} \sigma^{-(K-1)} (1-\rho)^{-\frac{K-1}{2}} \prod_{k=1}^{K-1} \left\{\frac{(1+T\xi + \lambda_k f^2)}{(1+\lambda_k f)^2}\right\}^{-1/2}$$

$$\cdot \exp\left\{-\frac{1}{2\sigma^2(1-\rho)}\left[\sum_{k=1}^{K-1} \frac{\hat{\eta}_k^2 (1+\lambda_k f)^2}{(1+T\xi+\lambda_k f^2)}\right]\right\} \tag{A.4}$$

where $\underline{\hat{\eta}} = P^{-1}[\underline{d}(\hat{\rho})-\underline{\delta}]$ and $\underline{\hat{\eta}} \equiv (\hat{\eta}_1,\hat{\eta}_2,\ldots,\hat{\eta}_{K-1})'$. We can find the exact finite sample distribution of $\underline{d}(\hat{\rho})$ by

$$p[\underline{d}(\hat{\rho})] = \int_0^\infty p[\underline{d}(\hat{\rho})|\, f]p(f)df \quad . \tag{A.5}$$

Analytical evaluation of the integral (A.5) does not appear possible. Following Mehta and Swamy (1970) we can use numerical methods to evaluate (A.5).

CHAPTER III

EFFICIENT METHODS OF ESTIMATING THE ERROR COMPONENTS REGRESSION MODELS

3.1 Introduction

We referred to a linear regression model combining cross-section with time series data in Section 1.2(b). This model contains some explanatory variables, a constant term, an additive disturbance term varying both in the time and cross-sectional dimensions, and two additive components--one component associated with time, another with the cross-sectional units. In this chapter we present an analysis of this model as an approach to combining cross-section and time series data.

In Section 3.2 we state some matrix results. In Section 3.3 we treat the additive components as fixed parameters and discuss an appropriate estimation procedure. In Section 3.4 we treat the components as random and develop efficient methods of estimation. In Section 3.5 we develop a class of asymptotically efficient estimators of coefficients. Section 3.6 is devoted to a study in the small sample properties of an estimator of coefficients developed in Section 3.4. In Sections 3.7 and 3.8 we compare the small sample efficiencies of different estimators of coefficients. In Section 3.9 we refer to some alternative estimators of coefficients and variances and to regressions on lagged values of the dependent variables. Estimators of coefficients when the intercept varies systematically either across units or over time are developed in Section 3.10. Maximum likelihood method of estimating error components models and the robustness of inferences based on an estimator of coefficients with respect to some departures in the underlying assumptions are considered in Sections 3.11 and 3.12. Concluding remarks are presented in Section 3.13.

3.2 Some Matrix Results

We state below some known matrix results and a definition of a generalized inverse of a rectangular or singular matrix.

Lemma 3.2.1: Let X be a TXK matrix of rank K and L be a TX(T-K) matrix of rank T-K such that $L'X=0$. Let Σ be a TXT nonsingular symmetric matrix. Then

$$(X'\Sigma^{-1}X)^{-1}X'\Sigma^{-1} = (X'X)^{-1}X' - (X'X)^{-1}X'\Sigma L(L'\Sigma L)^{-1}L' \quad .$$

The proof is given in Rao (1967, p. 358, Lemma 2b).

Lemma 3.2.2: If X is a TXK matrix of rank K and Σ is a TXT nonsingular symmetric matrix, then the matrix

$$(X'X)^{-1}X'\Sigma X(X'X)^{-1} - (X'\Sigma^{-1}X)^{-1}$$

is nonnegative definite.

The proof is given in Rao (1967, p. 358, Lemma 2c).

Definition 3.2.1: Let A be a m X n matrix of any rank. Generalized inverse of A is an n X m matrix, denoted by A^-, such that for any vector $\underline{y}$ for which $AX = \underline{y}$ is a consistent equation, $X = A^-\underline{y}$ is a solution. A^- exists if and only if $AA^-A = A$, cf. Rao (1966).

Generalized inverse as defined above is not necessarily unique.

3.3 Covariance Estimators

Combining all the nT observations we can write (1.2.6) as

$$\underline{y} = \beta_0 \underline{\iota}_{nT} + Z\underline{\delta} + \underline{u} \tag{3.3.1}$$

where $\underline{y} \equiv (y_{11},\ldots,y_{1T},\ldots,y_{n1},\ldots,y_{nT})'$ is a nT X 1 vector of observations on the dependent variable,

$$Z = \begin{bmatrix} z_{111} & \cdots & z_{k11} & \cdots & z_{K-1,11} \\ \vdots & & \vdots & & \vdots \\ z_{11T} & \cdots & z_{k1T} & \cdots & z_{K-1,1T} \\ \vdots & & \vdots & & \vdots \\ z_{1n1} & \cdots & z_{kn1} & \cdots & z_{K-1,n1} \\ \vdots & & \vdots & & \vdots \\ z_{1nT} & \cdots & z_{knT} & \cdots & z_{K-1,nT} \end{bmatrix}$$

is a nT X K$'$ matrix of observations on K-1 independent variables, β_0 is a constant, $\underline{\delta}$ is a K$'$X1 vector of slope coefficients, $u_{it} = \mu_i + \tau_t + \upsilon_{it}$, $\underline{u} \equiv (\mu_1 + \tau_1 + \upsilon_{11}, \ldots, \mu_1 + \tau_T + \upsilon_{1T}, \ldots, \mu_n + \tau_1 + \upsilon_{n1}, \ldots, \mu_n + \tau_T + \upsilon_{nT})'$ is a nTX1 vector of error components. The significance of μ_i, τ_t and υ_{it} is explained in (1.2.6). Let $X \equiv [\underline{\iota}_{nT}, Z]$ and $\underline{\beta} \equiv [\beta_0, \underline{\delta}']'$.

In this section we estimate the parameters of the model (3.3.1) under the following assumption for i&j=1,2, ..., n.

Assumption 3.3.1:

(1) Same as (1) in Assumption 2.6.1.

(2) X is fixed in repeated samples on $\underline{y}$. The rank of X is K.

(3) The μ_i's and τ_t's are fixed parameters obeying the restrictions $\sum_{i=1}^{n} \mu_i = 0$ and $\sum_{t=1}^{T} \tau_t = 0$.

(4) The vectors $\underline{\upsilon}_i \equiv (\upsilon_{i1}, \upsilon_{i2}, \ldots, \upsilon_{iT})'$ (i=1,2, ..., n) are independently and identically distributed with $E\underline{\upsilon}_i = 0$ and $E\underline{\upsilon}_i\underline{\upsilon}_i' = \sigma_\upsilon^2 I_T$.

Following Graybill (1961), Mundlak (1963), Wallace and Hussain (1969), and others we can sweep out the constants μ_i and τ_t for a given sample by the covariance transformation which can be represented by the matrix

$$Q = I_{nT} - T^{-1}(I_n \otimes \underline{\iota}_T\underline{\iota}_T') - n^{-1}(\underline{\iota}_n\underline{\iota}_n' \otimes I_T) + (nT)^{-1}\underline{\iota}_{nT}\underline{\iota}_{nT}' \quad . \tag{3.3.2}$$

Applying the convariance transformation to all the nT observations in (3.3.1) we have

$$Q\underline{y} = QX\underline{\beta} + Q\underline{u} = QZ\underline{\delta} + Q\underline{u} \quad . \tag{3.3.3}$$

The second equality in (3.3.3) follows from the fact that $Q\underline{\iota}_{nT} = 0$. Using Assumption 3.3.1 we can show that $EQ\underline{u} = 0$ and $EQ\underline{u}\underline{u}'Q = \sigma_\upsilon^2 Q$ which is singular. Consequently, the usual Aitken estimator of $\underline{\delta}$ does not exist. Since the variance-covariance matrix of $Q\underline{u}$ is not scalar, application of OLS might lead to an inefficient estimator of $\underline{\delta}$. Wallace and Hussain apply OLS to (3.3.3) without providing proper justification. It has been shown by Mitra and Rao (1968b) that an unconstrained Aitken estimator based on any generalized inverse of Q is the BLUE of $\underline{\delta}$ since the linear subspace spanned by the columns of QZ is contained in the linear subspace spanned by the columns of the variance-covariance matrix of the disturbance vector $Q\underline{u}$. Since Q is an idempotent matrix, QQQ=Q. This shows that according to Definition 3.2.1 a generalized inverse of an idempotent matrix is itself. The Aitken estimator of $\underline{\delta}$ based on a generalized inverse of Q is

$$\tilde{\underline{\delta}} = (Z'QZ)^{-1}Z'Q\underline{y} \quad . \tag{3.3.4}$$

The estimator $\tilde{\underline{\delta}}$ is the same as an estimator obtained by applying OLS to the equation (3.3.3). If Assumption 3.3.1 is true, then $\tilde{\underline{\delta}}$ is the BLUE of $\underline{\delta}$. In actual computation it is convenient first, to take the mean of time series observations separately for each unit, to take the mean of cross-section observations separately for each year, to take the overall mean of all observations and next, to transform the observations by subtracting from each observation the corresponding time mean and cross-section mean and adding the overall mean. If we apply OLS to these transformed observations we obtain (3.3.4).

3.4 Estimation of Error Components Models

(a) Aitken Estimator Based on Estimated Variances of Disturbances

In this section we estimate the model (3.3.1) under the following assumption for i&j=1,2, ..., n and t&t′ = 1,2, ..., T.

Assumption 3.4.1:

(1) Same as (1) in Assumption 2.6.1.

(2) Same as (2) in Assumption 3.3.1.

(3) Same as (3) in Assumption 2.4.1.

(4) $E\tau_t=0$ for all t and $E\tau_t\tau_{t'} = \begin{cases}\sigma_\tau^2 & \text{if } t=t'; \\ 0 & \text{if } t \neq t'.\end{cases}$

(5) Same as (4) in Assumption 2.4.1.

(6) μ_i, τ_t and υ_{it} are independent of each other.

Under Assumption 3.4.1 the variance-covariance matrix of $\underline{u}$ in (3.3.1) is

$$E\underline{u}\underline{u}' = \sigma_\mu^2 (I_n \otimes \iota_T\iota_T') + \sigma_\tau^2 (\iota_n\iota_n' \otimes I_T) + \sigma_\upsilon^2 I_{nT} \quad . \tag{3.4.1}$$

We can compare the structure of (3.4.1) with that of (2.4.3). The presence of separate random year effects τ_t introduces correlations among the disturbance terms of different individuals.

It has been noted by Nerlove (1967) that we obtain the same form of variance-covariance matrix as in (3.4.1) for $\underline{u}$ if we set $\tau_t \equiv 0$ for all t and add the assumption that $E(\mu_i\upsilon_{it} + \mu_j\upsilon_{it'}) = 2\sigma_{\mu\upsilon}$ if $t=t'$ for all i&j, 0 otherwise. In this case $Eu_{it}u_{jt}$ for $i \neq j$ can be negative.

Now consider the orthogonal matrix O_T in (2.3.5). Applying the transformation defined by $Q_1 = (I_n \otimes \iota_T'/\sqrt{T})$ to all the nT observations in (3.3.1) we have

$$\underline{y}_1 = X_1\underline{\beta} + \underline{u}_1 \tag{3.4.2}$$

where $\underline{y}_1 = Q_1\underline{y}$ is a nX1 vector, $X_1 = Q_1X$ is a nXK matrix and $\underline{u}_1 = Q_1\underline{u}$ is a nX1 vector of transformed disturbances. We can easily recognize that the observations in (3.4.2) are proportional to the means of time series observations on different variables for different units. For the mean and the variance-covariance matrix of $\underline{u}_1$ we have

$$\begin{aligned} E\underline{u}_1 &= 0 \text{ and} \\ E\underline{u}_1\underline{u}_1' &= EQ_1\underline{u}\underline{u}'Q_1' \\ &= (I_n \otimes \iota_T'/\sqrt{T})[\sigma_\mu^2(I_n \otimes \iota_T\iota_T') + \sigma_\tau^2(\iota_n\iota_n' \otimes I_T) \\ &\qquad + \sigma_\upsilon^2 I_{nT}](I_n \otimes \iota_T/\sqrt{T}) \\ &= T\sigma_\mu^2 I_n + \sigma_\tau^2 \iota_n\iota_n' + \sigma_\upsilon^2 I_n \\ &= \sigma_1^2 [\rho_1 \iota_n\iota_n' + (1-\rho_1)I_n] \end{aligned} \tag{3.4.3}$$

where $\sigma_1^2 = T\sigma_\mu^2 + \sigma_\tau^2 + \sigma_\upsilon^2$, $\rho_1 = \sigma_\tau^2/\sigma_1^2$.[15] Clearly, $0 \le \rho_1 \le 1$. The variance-covariance matrix of $\underline{u}_1$ is singular if $\rho_1 = 1$. Therefore we assume that $0 \le \rho_1 < 1$.

---

[15] Here we utilized the result $(A \otimes B)(C \otimes D) = AC \otimes BD$. A collection of properties of Kronecker products and some applications were provided by Neudecker (1969).

Due to the special form of the variance-covariance matrix of $\underline{u}_1$ and the presence of the column $\sqrt{T}\underline{\iota}_n$ in X_1, it follows from the condition (2.3.4a) that the BLUE of $\underline{\beta}$ in the setup (3.4.2) is

$$\hat{\underline{\beta}}(1) = (X_1'X_1)^{-1}X_1'\underline{y}_1 \quad . \tag{3.4.4}$$

The subvector of $\hat{\underline{\beta}}(1)$ corresponding to slope coefficients is

$$\hat{\underline{\delta}}(1) = (Z_1'N_1Z_1)^{-1}Z_1'N_1\underline{y}_1 \tag{3.4.5}$$

where $Z_1 = Q_1Z$ is a nXK' matrix and $N_1 = I_n - \underline{\iota}_n\underline{\iota}_n'(1/n)$. The subscript 1 is attached to the estimator (3.4.5) only to distinguish it from the other estimators of $\underline{\delta}$ to be developed later in this subsection. The variance-covariance matrix of $\hat{\underline{\delta}}(1)$ is

$$V[\hat{\delta}(1)] = \sigma_1^{\,2}(1-\rho_1)(Z_1'N_1Z_1)^{-1} \quad . \tag{3.4.6}$$

An unbiased estimator of $\sigma_1^{\,2}(1-\rho_1) = \bar{\sigma}_1^{\,2} = (T\sigma_\mu^{\,2} + \sigma_\upsilon^{\,2})$ is

$$\hat{\bar{\sigma}}_1^{\,2} = \underline{y}_1'M_1\underline{y}_1/(n-K) \tag{3.4.7}$$

where $M_1 = I_n - X_1(X_1'X_1)^{-1}X_1'$.

Now consider an orthogonal matrix, say O_n, of order n such that its first row is equal to $\underline{\iota}_n'/\sqrt{n}$. Let $O_n = [\underline{\iota}_n/\sqrt{n}, C_2']'$. Because O_n is orthogonal, we have $C_2\underline{\iota}_n = 0$, $C_2'C_2 = I_n - \underline{\iota}_n\underline{\iota}_n'(1/n) = N_1$, $C_2C_2' = I_{n'}$, $n' = n-1$. Applying the transformation defined by $Q_2 = (\underline{\iota}_n'/\sqrt{n} \otimes I_{T'}) \cdot (I_n \otimes C_1) = (\underline{\iota}_n'/\sqrt{n} \otimes C_1)$ to all the nT observations in (3.3.1) we have

$$\underline{y}_2 = Z_2\underline{\delta} + \underline{u}_2 \tag{3.4.8}$$

where $\underline{y}_2 = Q_2\underline{y}$ is a $T'X1$ vector of transformed observations on the dependent variable, $Z_2 = Q_2Z$ is a $T'XK'$ matrix of transformed observations on K' independent variables, $\underline{u}_2 = Q_2\underline{u}$ is a $T'X1$ vector of transformed disturbances, and use is made of the result $Q_2\underline{\iota}_{nT} = 0$. The mean and the variance-covariance matrix of $\underline{u}_2$ is straightforwardly derived as $E\underline{u}_2 = 0$ and

$$\begin{aligned}E\underline{u}_2\underline{u}_2' &= EQ_2\underline{u}\underline{u}'Q_2' \\ &= (\underline{\iota}_n'/\sqrt{n} \otimes C_1)[\sigma_\mu^2(I_n \otimes \underline{\iota}_T\underline{\iota}_T') + \sigma_\tau^2(\underline{\iota}_n\underline{\iota}_n' \otimes I_T) + \sigma_\upsilon^2 I_{nT}]\cdot(\underline{\iota}_n/\sqrt{n} \otimes C_1') \\ &= (n\sigma_\tau^2 + \sigma_\upsilon^2)I_{T'} \quad . \end{aligned} \tag{3.4.9}$$

The third equality in (3.4.9) follows from the results $C_1\underline{\iota}_T = 0$ and $C_1C_1' = I_{T'}$. Let $\sigma_2^2 = \sigma_\tau^2 + \sigma_\upsilon^2$, $\rho_2 = \sigma_\tau^2/\sigma_2^2$ and $\bar{\sigma}_2^2 = n\sigma_\tau^2 + \sigma_\upsilon^2 = \sigma_2^2(1 - \rho_2 + n\rho_2)$. We assume that $0 \le \rho_2 < 1$.

Applying OLS to (3.4.8) we obtain

$$\hat{\underline{\delta}}(2) = (Z_2'Z_2)^{-1}Z_2'\underline{y}_2 \quad . \tag{3.4.10}$$

The variance-covariance matrix of $\hat{\underline{\delta}}(2)$ is $\bar{\sigma}_2^2(Z_2'Z_2)^{-1}$. An unbiased estimator of $\bar{\sigma}_2^2$ is

$$\hat{\bar{\sigma}}_2^2 = \underline{y}_2'M_2\underline{y}_2/(T-K) \tag{3.4.11}$$

where $M_2 = I_{T'} - Z_2(Z_2'Z_2)^{-1}Z_2'$.

Finally, applying the transformation defined by $Q_3 = (C_2 \otimes I_{T'})(I_n \otimes C_1) = (C_2 \otimes C_1)$ to all the nT observations in (3.3.1) we have

$$\underline{y}_3 = Z_3\underline{\delta} + \underline{u}_3 \tag{3.4.12}$$

where $\underline{y}_3 = Q_3\underline{y}$ is a $n'T'X1$ vector of transformed observations on the dependent variable, $Z_3 = Q_3Z$ is a $n'T'XK'$ matrix of transformed observations on K' independent variables, and $\underline{u}_3 = Q_3\underline{u}$ is a $n'T'X1$ vector of transformed disturbances. It can be shown that $E\underline{u}_3 = 0$ and

$$\begin{aligned}E\underline{u}_3\underline{u}_3' &= Q_3E\underline{u}\underline{u}'Q_3' \\ &= (C_2 \otimes C_1)[\sigma_\mu^2(I_n \otimes \underline{\iota}_T\underline{\iota}_T') + \sigma_\tau^2(\underline{\iota}_n\underline{\iota}_n' \otimes I_T) + \sigma_\upsilon^2 I_{nT}]\cdot(C_2' \otimes C_1') \\ &= \sigma_\upsilon^2 I_{n'T'} \end{aligned} \tag{3.4.13}$$

where use is made of the results $C_1\underline{\iota}_T = 0$ and $C_2\underline{\iota}_n = 0$.

OLS applied to (3.4.12) gives the estimator

$$\hat{\underline{\delta}}(3) = (Z_3'Z_3)^{-1}Z_3'\underline{y}_3 \quad . \tag{3.4.14}$$

The variance-covariance matrix of $\underline{\hat{\delta}}(3)$ is $\sigma_\upsilon^2(Z_3Z_3)^{-1}$ and an unbiased estimator of σ_υ^2 is

$$\hat{\sigma}_\upsilon^2 = \underline{y}_3' M_3 \underline{y}_3/(n'T' - K + 1) \tag{3.4.15}$$

where $M_3 = I_{n'T'} - Z_3(Z_3'Z_3)^{-1}Z_3'$.

Notice that $Q_1Q_2' = 0, Q_2Q_3' = 0$. The rank of Q_1 is equal to the rank of I_n multiplied by the rank of $\underline{\iota}_T'/\sqrt{T}$ because if A and B are any arbitrary matrices, then the rank of $(A \otimes B)$ is equal to the rank of A multiplied by the rank of B. Therefore rank (Q_1) + rank (Q_2) + rank (Q_3) = $n + T' + n'T'$ = nT which is the total number of observations. This indicates that in estimating (3.4.2), (3.4.8) and (3.4.12) we used up all the orthogonal linear combinations of the available observations. From (3.4.5), (3.4.10) and (3.4.14) we have three uncorrelated estimators of the same parametric vector and we can pool them in the following manner.

$$\underline{\hat{\delta}}(\underline{f}) = \left[\frac{Z_1'N_1Z_1}{\hat{\bar{\sigma}}_1^2} + \frac{Z_2'Z_2}{\hat{\bar{\sigma}}_2^2} + \frac{Z_3'Z_3}{\hat{\sigma}_\upsilon^2}\right]^{-1}\left[\frac{Z_1'N_1\underline{y}_1}{\hat{\bar{\sigma}}_1^2} + \frac{Z_2'\underline{y}_2}{\hat{\bar{\sigma}}_2^2} + \frac{Z_3'\underline{y}_3}{\hat{\sigma}_\upsilon^2}\right] \tag{3.4.16}$$

where $\underline{f} = (f_1, f_2)'$, $f_1 = \hat{\bar{\sigma}}_1^2/\hat{\sigma}_\upsilon^2$ and $f_2 = \hat{\bar{\sigma}}_1^2/\hat{\bar{\sigma}}_2^2$. In a compact form we write

$$\underline{\hat{\delta}}(\underline{f}) = (W'S_d^{-1}W)^{-1}W'S_d^{-1}\underline{r} \tag{3.4.17}$$

where $W = [Z_1'N_1, Z_2', Z_3']'$ is a nTXK′ matrix, $S_d \equiv \text{diag}\,[\hat{\bar{\sigma}}_1^2 I_n, \hat{\bar{\sigma}}_2^2 I_{T'}, \hat{\sigma}_\upsilon^2 I_{n'T'}]$ and $\underline{r} \equiv [\underline{y}_1'N_1, \underline{y}_2', \underline{y}_3']'$.

We can easily recognize that $\underline{\hat{\delta}}(\underline{f})$ is an Aitken estimator of $\underline{\delta}$ based on estimated variances of disturbances. One obvious weakness of Aitken procedure is that it is unable to utilize the restrictions $\bar{\sigma}_1^2 \geq \sigma_\upsilon^2$ and $\bar{\sigma}_2^2 \geq \sigma_\upsilon^2$ implied by the reparameterized version of the model. If the values taken by the estimators $\hat{\bar{\sigma}}_1^2$ and $\hat{\bar{\sigma}}_2^2$ in a given sample are appreciably smaller than the value taken by $\hat{\sigma}_\upsilon^2$, then the specifications in (3.3.1) and (3.4.1) may contain some errors. An Aitken estimator of $\underline{\delta}$ is

$$\underline{\hat{\delta}}(\underline{c}) = \left[\frac{Z_1'N_1Z_1}{\bar{\sigma}_1^2} + \frac{Z_2'Z_2}{\bar{\sigma}_2^2} + \frac{Z_3'Z_3}{\sigma_\upsilon^2}\right]^{-1}\left[\frac{Z_1'N_1\underline{y}_1}{\bar{\sigma}_1^2} + \frac{Z_2'\underline{y}_2}{\bar{\sigma}_2^2} + \frac{Z_3'\underline{y}_3}{\sigma_\upsilon^2}\right] \tag{3.4.18}$$

where $\underline{c} = (c_1, c_2)'$, $c_1 = \bar{\sigma}_1^2/\sigma_\upsilon^2$ and $c_2 = \bar{\sigma}_2^2/\sigma_\upsilon^2$.

Wallace and Hussain (1969) have also developed an Aitken estimator of $\underline{\delta}$ using some consistent estimators of variance components σ_μ^2, σ_τ^2 and σ_υ^2. Their variance estimators are different from the estimators of σ_μ^2, σ_τ^2 and σ_υ^2 implied by (3.4.7), (3.4.11) and (3.4.15), and are based on observed residuals obtained by OLS, applied directly to (3.3.1). It will be clear from our subsequent discussion that the procedure outlined above is convenient to derive the small samples properties of an Aitken estimator of $\underline{\delta}$ based on estimated variances of errors. We will also show that under very general conditions, $\hat{\underline{\delta}}(\underline{f})$ is consistent even when lagged values of the dependent variable appear as explanatory variables in (3.3.1). The estimators developed by Wallace and Hussain are not consistent in this case.

Notice that the estimator $\tilde{\underline{\delta}}$ in (3.3.4) is the same as the estimator $\hat{\underline{\delta}}(3)$ in (3.4.14) because $Q = Q_3'Q_3$. If we apply OLS to (3.3.1) we obtain the following estimator for the slope coefficients.

$$\begin{aligned}\hat{\underline{\delta}} &= (Z'N\,Z)^{-1}Z'N\underline{y} \\ &= (W'W)^{-1}W'\underline{r}\end{aligned} \tag{3.4.19}$$

where $N = I_{nT} - (nT)^{-1}\underline{\iota}_{nT}\underline{\iota}_{nT}'$ and the second equality is based on the results $Q_1'N_1Q_1 = (I_n \otimes \underline{\iota}_T/\sqrt{T})(I_n - \underline{\iota}_n\underline{\iota}_n'/n)(I_n \otimes \underline{\iota}_T'/\sqrt{T}) = (I_n \otimes \underline{\iota}_T\underline{\iota}_T'T^{-1}) - (nT)^{-1}\underline{\iota}_{nT}\underline{\iota}_{nT}'$, $Q_2'Q_2 = (n^{-1}\underline{\iota}_n\underline{\iota}_n' \otimes I_T) - (nT)^{-1}\underline{\iota}_{nT}\underline{\iota}_{nT}'$, $Q = Q_3'Q_3$ and $N = Q_1'N_1Q_1 + Q_2'Q_2 + Q$.

(b) Computational Procedure

The procedures of computing $\hat{\underline{\delta}}(1)$ and $\hat{\underline{\delta}}(3)$ are already discussed in subsection 2.4(b) and Section 3.3 respectively. Now we explain how to compute $\hat{\underline{\delta}}(2)$. We write

$$\begin{aligned}\hat{\underline{\delta}}(2) &= (Z'Q_2'Q_2Z)^{-1}Z'Q_2'Q_2\underline{y} \\ &= [Z'(n^{-1}\underline{\iota}_n\underline{\iota}_n' \otimes I_T)Z - (nT)^{-1}Z'\underline{\iota}_{nT}\underline{\iota}_{nT}'Z]^{-1} \\ &\quad \cdot [Z'(n^{-1}\underline{\iota}_n\underline{\iota}_n' \otimes I_T)\underline{y} - (nT)^{-1}Z'\underline{\iota}_{nT}\underline{\iota}_{nT}'\underline{y}] \,.\end{aligned}$$

By taking the mean of time series observations separately for each unit we obtain n means. We express the time series observations on each variable for each unit as deviations from their own mean . By averaging these deviations across units separately for each year we obtain T averages for each variable. If z_{kit} ($i=1,2,\ldots,n$; $t=1,2,\ldots T$) represent nT observations on the k-th independent variable, then the T averages for this variable are given by $n^{-1}\sum_{i=1}^{n}(z_{kit} - T^{-1}\sum_{t=1}^{T} z_{kit})$ ($t=1,2,\ldots,T$). Taking these T averages as observations if we apply OLS we obtain $\hat{\underline{\delta}}(2)$. Thus, we do not have to construct the matrices C_1 and C_2 before computing $\hat{\underline{\delta}}(2)$.

After computing $\hat{\underline{\delta}}(1)$, $\hat{\underline{\delta}}(2)$ and $\hat{\underline{\delta}}(3)$ we can pool them using the formula (3.4.16).

3.5 A Class of Asymptotically Efficient Estimators

We define an $\underline{\alpha}$-II class of estimators which covers the estimators $\hat{\underline{\delta}}(\underline{f})$, $\tilde{\underline{\delta}}$, $\hat{\underline{\delta}}$, $\hat{\underline{\delta}}(1)$ and $\hat{\underline{\delta}}(2)$ as its members.

$$\hat{\underline{\delta}}(\underline{f},\underline{\alpha}) = \left[\frac{\alpha_0 Z_1' N_1 Z_1}{\alpha_1\hat{\bar{\sigma}}_1^2 + \alpha_2\hat{\sigma}_\upsilon^2} + \frac{\alpha_3 Z_2' Z_2}{\alpha_4\hat{\bar{\sigma}}_2^2 + \alpha_5\hat{\sigma}_\upsilon^2} + \frac{\alpha_6 Z_3' Z_3}{\hat{\sigma}_\upsilon^2}\right]^{-1} \cdot \left[\frac{\alpha_0 Z_1' N_1 \underline{y}_1}{\alpha_1\hat{\bar{\sigma}}_1^2 + \alpha_2\hat{\sigma}_\upsilon^2} + \frac{\alpha_3 Z_2' \underline{y}_2}{\alpha_4\hat{\bar{\sigma}}_2^2 + \alpha_5\hat{\sigma}_\upsilon^2} + \frac{\alpha_6 Z_3' \underline{y}_3}{\hat{\sigma}_\upsilon^2}\right] \tag{3.5.1}$$

where $\underline{\alpha} = (\alpha_0,\alpha_1,\alpha_2,\alpha_3,\alpha_4,\alpha_5,\alpha_6)'$, the α_i's are nonnegative constants obeying the restriction that only one of α_1 and α_2, and only one of α_4 and α_5 can be equal to zero at any one time. It is obvious that $\hat{\underline{\delta}}(\underline{f})$ in (3.4.16) is an $\underline{\alpha}$-II estimator with $\underline{\alpha} = (1,1,0,1,1,0,1)'$ or with ($\alpha_0 = \alpha_3 = \alpha_6 \neq 0, \alpha_1 = \alpha_4 = 1$, $\alpha_2 = \alpha_5 = 0$) or with ($\alpha_2 = \alpha_5 = 0$, $\alpha_6 = 1$, $\alpha_0 = \alpha_1$ and $\alpha_3 = \alpha_4$). $\hat{\underline{\delta}}$ in (3.4.19) is an $\underline{\alpha}$-II estimator with $\underline{\alpha} = (1,0,1,1,0,1,1)'$ or with ($\alpha_1 = \alpha_4 = 0$, $\alpha_2 = \alpha_5 = 1$ and $\alpha_0 = \alpha_3 = \alpha_6 \neq 0$).* $\tilde{\underline{\delta}}$ in (3.3.4) or $\hat{\underline{\delta}}(3)$ in (3.4.14) is an $\underline{\alpha}$-II estimator with $\underline{\alpha} = (0,\alpha_1,\alpha_2,0,\alpha_4,\alpha_5,\alpha_6 \neq 0)'$. $\hat{\underline{\delta}}(1)$ in (3.4.5) is an $\underline{\alpha}$-II estimator with $\underline{\alpha} = (\alpha_0 \neq 0,\alpha_1,\alpha_2,0,\alpha_4,\alpha_5,0)'$ and $\hat{\underline{\delta}}(2)$ in (3.4.10) is an $\underline{\alpha}$-II estimator with $\underline{\alpha} = (0,\alpha_1,\alpha_2,\alpha_3 \neq 0,\alpha_4,\alpha_5,0)'$.

Under the assumption that $\lim_{\substack{n\to\infty\\T\to\infty}} X'X/nT$ is a finite, positive definite matrix and the X's are weakly nonstochastic (do not repeat in repeated samples) Wallace

---

* $\hat{\underline{\delta}}$ is also an $\underline{\alpha}$-II estimator with ($\alpha_1 = \alpha_4 = 0$, $\alpha_0 = \alpha_2, \alpha_3 = \alpha_5, \alpha_6 = 1)'$.

and Hussain (1969) have shown that the Aitken estimator $\hat{\underline{\delta}}(\underline{c})$, an Aitken estimator based on estimated variance components and the estimator $\tilde{\underline{\delta}}$ have the same asymptotic variance-covariance matrix. To establish the asymptotic properties of $\hat{\underline{\delta}}(\underline{f})$ we also make use of similar assumptions.

Assumption 3.5.1:

(1) The X's are weakly nonstochastic

(2) $\lim_{\substack{n\to\infty\\T\to\infty}}(nT)^{-1}Z_1'N_1Z_1$, $\lim_{\substack{n\to\infty\\T\to\infty}}(nT)^{-1}Z_2'Z_2$ and $\lim_{\substack{n\to\infty\\T\to\infty}}(nT)^{-1}Z_3'Z_3$ are all finite and positive definite matrices.

Assumption 3.5.2: The disturbance vector $\underline{u}$ is normally distributed.

It is easy to show that

$$\operatorname*{plim}_{\substack{n\to\infty\\T\to\infty}} \hat{\bar{\sigma}}_1^{\,2}/T=\sigma_\mu^2;\quad \operatorname*{plim}_{\substack{n\to\infty\\T\to\infty}} \hat{\bar{\sigma}}_2^{\,2}/n=\sigma_\tau^2 \text{ and } \operatorname*{plim}_{\substack{n\to\infty\\T\to\infty}} \hat{\sigma}_\upsilon^2=\sigma_\upsilon^2 \quad . \tag{3.5.2}$$

We can write

$$\begin{aligned}\sqrt{nT}[\hat{\underline{\delta}}(\underline{f},\underline{\alpha})-\underline{\delta}] = {} & \left[\frac{\alpha_0 Z_1'N_1Z_1/nT^2}{(\alpha_1\hat{\bar{\sigma}}_1^{\,2}+\alpha_2\hat{\sigma}_\upsilon^2)/T} + \frac{\alpha_3 Z_2'Z_2/n^2T}{(\alpha_4\hat{\bar{\sigma}}_2^{\,2}+\alpha_5\hat{\sigma}_\upsilon^2)/n}\right.\\ & \left.+\frac{\alpha_6 Z_3'Z_3/nT}{\hat{\sigma}_\upsilon^2}\right]^{-1}\left[\frac{\alpha_0 Z_1'N_1\underline{u}_1/T\sqrt{nT}}{(\alpha_1\hat{\bar{\sigma}}_1^{\,2}+\alpha_2\hat{\sigma}_\upsilon^2)/T}\right.\\ & \left.+\frac{\alpha_3 Z_2'\underline{u}_2/n\sqrt{nT}}{(\alpha_4\hat{\bar{\sigma}}_2^{\,2}+\alpha_5\hat{\sigma}_\upsilon^2)/n}+\frac{\alpha_6 Z_3'\underline{u}_3/\sqrt{nT}}{\hat{\sigma}_\upsilon^2}\right] .\end{aligned} \tag{3.5.3}$$

It follows from (3.5.2) that $\operatorname*{plim}_{\substack{n\to\infty\\T\to\infty}} (\alpha_1\hat{\bar{\sigma}}_1^{\,2} + \alpha_2\hat{\sigma}_\upsilon^2)/T = \alpha_1\sigma_\mu^2$ and $\operatorname*{plim}_{\substack{n\to\infty\\T\to\infty}} (\alpha_4\hat{\bar{\sigma}}_2^{\,2} + \alpha_5\hat{\sigma}_\upsilon^2)/n = \alpha_4\sigma_\tau^2$. Also notice that $\lim_{\substack{n\to\infty\\T\to\infty}} Z_1'N_1Z_1/nT^2 = 0$ and $\lim_{\substack{n\to\infty\\T\to\infty}} Z_2'Z_2/n^2T = 0$.

Consequently, when $\alpha_1 \neq 0$, $\alpha_4 \neq 0$,

$$\operatorname*{plim}_{\substack{n\to\infty\\T\to\infty}}\left[\frac{\alpha_0 Z_1'N_1Z_1/nT^2}{(\alpha_1\hat{\bar{\sigma}}_1^{\,2}+\alpha_2\hat{\sigma}_\upsilon^2)/T} + \frac{\alpha_3 Z_2'Z_2/n^2T}{(\alpha_4\hat{\bar{\sigma}}_2^{\,2}+\alpha_5\hat{\sigma}_\upsilon^2)/n} + \frac{\alpha_6 Z_3'Z_3/nT}{\hat{\sigma}_\upsilon^2}\right] \tag{3.5.4}$$

is the same as $\operatorname*{plim}_{\substack{n\to\infty \\ T\to\infty}} \alpha_6 Z_3' Z_3 / nT\hat{\sigma}_\upsilon^2$ which is finite and positive definite according to Assumption 3.5.1 provided $\alpha_6 \neq 0$. We conclude from (3.5.3) that the asymptotic distribution of $\sqrt{nT}[\hat{\underline{\delta}}(\underline{f},\underline{\alpha})-\underline{\delta}]$ is the same as that of the inverse of limit matrix (3.5.4) postmultiplied by

$$\left[\frac{\alpha_0 Z_1' N_1 \underline{u}_1 / T\sqrt{nT}}{(\alpha_1 \hat{\bar{\sigma}}_1^2 + \alpha_2 \hat{\sigma}_\upsilon^2)/T} + \frac{\alpha_3 Z_2' \underline{u}_2 / n\sqrt{nT}}{(\alpha_4 \hat{\bar{\sigma}}_2^2 + \alpha_5 \hat{\sigma}_\upsilon^2)/n} + \frac{\alpha_6 Z_3' \underline{u}_3 / \sqrt{nT}}{\hat{\sigma}_\upsilon^2}\right] . \tag{3.5.5}$$

Since the first two terms in (3.5.5) converge in probability to zero,

$$\operatorname*{plim}_{\substack{n\to\infty \\ T\to\infty}} \sqrt{nT}\,|\hat{\underline{\delta}}(\underline{f},\underline{\alpha})-\tilde{\underline{\delta}}| = 0 . \tag{3.5.6}$$

Consequently, according to the limit theorem (ix) in Rao (1965a, p. 101) both $\hat{\underline{\delta}}(\underline{f},\underline{\alpha})$ and $\tilde{\underline{\delta}}$ have the same limiting distribution provided $\alpha_1 \neq 0$, $\alpha_4 \neq 0$ and $\alpha_6 \neq 0$. This rules out the possibility that the OLS estimator $\hat{\underline{\delta}}$ is asymptotically as efficient as $\tilde{\underline{\delta}}$. In fact, Wallace and Hussain (1969) have shown that the asymptotic variances of the elements of $\hat{\underline{\delta}}$ are unbounded. There are also other members of $\underline{\alpha}$-II class which are asymptotically inefficient. For example, consider $\underline{\alpha}$-II estimator with $\underline{\alpha} = (0,\alpha_1,\alpha_2,\alpha_3 \neq 0,0,\alpha_5,\alpha_6 \neq 0)'$. The variance-covariance matrix of $\sqrt{nT}$ times this estimator is

$$\left[\alpha_3 \frac{Z_2' Z_2}{\alpha_5 nT} + \alpha_6 \frac{Z_3' Z_3}{nT}\right]^{-1} \left[\bar{\sigma}_2^2 \alpha_3^2 \frac{Z_2' Z_2}{\alpha_5^2 nT} + \sigma_\upsilon^2 \alpha_6^2 \frac{Z_3' Z_3}{nT}\right] \left[\alpha_3 \frac{Z_2' Z_2}{\alpha_5 nT} + \alpha_6 \frac{Z_3' Z_3}{nT}\right]^{-1} .$$

The diagonal elements of this matrix tends to ∞ as $n\to\infty$, $T\to\infty$ because $\bar{\sigma}_2^2 \to \infty$ as $n\to\infty$. Similarly, the asymptotic variances of the elements of $\sqrt{nT}[\hat{\underline{\delta}}(\underline{f},\underline{\alpha})-\underline{\delta}]$ are unbounded if $\underline{\alpha}$ takes any one of the following values:

$$\underline{\alpha} = (0,\alpha_1,\alpha_2,\alpha_3 \neq 0,\alpha_4,\alpha_5,0)', \quad \underline{\alpha} = (\alpha_0 \neq 0,0,\alpha_2,0,\alpha_4,\alpha_5,\alpha_6 \neq 0)' ,$$

$$\underline{\alpha} = (\alpha_0 \neq 0,\alpha_1,\alpha_2,0,\alpha_4,\alpha_5,0)', \quad \underline{\alpha} = (\alpha_0 \neq 0,0,\alpha_2,\alpha_3 \neq 0,0,\alpha_5,\alpha_6 = 0)' ,$$

$$\underline{\alpha} = (\alpha_0 \neq 0,\alpha_1 = \alpha_4 = \alpha_6 = 0,\alpha_3 \neq 0,\alpha_2,\alpha_5)' .$$

These show that $\hat{\underline{\delta}}(1)$ and $\hat{\underline{\delta}}(2)$ are asymptotically inefficient.[16]

On the other hand, the results in (3.5.4)-(3.5.6) indicate that every $\underline{\alpha}$-II estimator with $\underline{\alpha} = (\alpha_0, \alpha_1 \neq 0, \alpha_2, \alpha_3, \alpha_4 \neq 0, \alpha_5, \alpha_6 \neq 0)'$ has an asymptotic distribution which is the same as that of $\hat{\underline{\delta}}(\underline{c})$. The point to be made in this subsection is that the set of asymptotically efficient estimators of $\underline{\delta}$ is infinite. It is not possible to choose an estimator of $\underline{\delta}$ on the basis of asymptotic efficiency. The small sample properties of different estimators of $\underline{\delta}$, to which we turn in the next section, provide a basis for choosing a feasible efficient estimator of $\underline{\delta}$.

3.6 Small Sample Properties of the Pooled Estimator

Writing $\underline{r} = W\underline{\delta} + \underline{v}$, where $\underline{v} = [\underline{u}_1' N_1, \underline{u}_2', \underline{u}_3']'$, we have

$$\hat{\underline{\delta}}(\underline{f}) = \underline{\delta} + (W' S_d^{-1} W)^{-1} W' S_d^{-1} \underline{v} \quad . \tag{3.6.1}$$

Under Assumption 3.5.2, $Z_1' N_1 \underline{u}_1$ is independent of $\hat{\bar{\sigma}}_1^2$, $Z_2' \underline{u}_2$ is independent of $\hat{\bar{\sigma}}_2^2$, and $Z_3' \underline{u}_3$ is independent of $\hat{\sigma}_\upsilon^2$ because $M_1 X_1 = 0$, $M_2 Z_2 = 0$, and $M_3 Z_3 = 0$. Furthermore, $\underline{u}_1, \underline{u}_2, \underline{u}_3$ are independent of each other. Consequently,

$$E\,[\hat{\underline{\delta}}(\underline{f}) \,|\, \underline{f}] = \underline{\delta} \quad \text{and} \quad E_{\underline{f}} \underline{\delta} = \underline{\delta} \quad . \tag{3.6.2}$$

This proves that $\hat{\underline{\delta}}(\underline{f})$ is an unbiased estimator of $\underline{\delta}$. In the same way we can show that all the members of $\underline{\alpha}$-II class in (3.5.1) are unbiased estimators of $\underline{\delta}$. The variance-covariance matrix of $\hat{\underline{\delta}}(\underline{f})$ conditional on S_d is

$$(W' S_d^{-1} W)^{-1} W' S_d^{-1} \Sigma_d S_d^{-1} W (W' S_d^{-1} W)^{-1} \tag{3.6.3}$$

where $\Sigma_d \equiv \text{diag}[\bar{\sigma}_1^2 I_n, \bar{\sigma}_2^2 I_{T'}, \sigma_\upsilon^2 I_{n'T'}]$. Let $\Sigma_d^{1/2} = \text{diag}[\bar{\sigma}_1 I_n, \bar{\sigma}_2 I_{T'}, \sigma_\upsilon I_{n'T'}]$. Since $W' \Sigma_d^{-1} W$ is a symmetric positive definite matrix, we can find an orthogonal matrix, say $O_{K'}$, of order $K' = K-1$, such that $O_{K'}' W' \Sigma_d^{-1} W O_{K'} = D^2$, a diagonal matrix with positive diagonal elements. Let $C_3 = D^{-1} O_{K'}' W' \Sigma_d^{-1/2}$. It immediately follows

---

[16] Notice, however, that $\sqrt{n}[\hat{\underline{\delta}}(1) - \underline{\delta}]$ with $\underline{\alpha} = (\alpha_0 \neq 0, \alpha_1, \alpha_2, 0, \alpha_4, \alpha_5, 0)'$ and $\sqrt{T}[\hat{\underline{\delta}}(2) - \underline{\delta}]$ with $\underline{\alpha} = (0, \alpha_1, \alpha_2, \alpha_3 \neq 0, \alpha_4, \alpha_5, 0)'$ have finite asymptotic variance-covariance matrices.

that $C_3C_3'=I_{K'}$. We can find a matrix C_4 of order $(nT-K')XnT$ such that $C_4C_4' = I_{nT-K'}$, $C_3'C_3 = I_{nT} - C_4'C_4$ and $[C_3', C_4']$ is an orthogonal matrix of order nT. Let $S_d^* = \Sigma_d^{-1/2} S_d \Sigma_d^{-1/2}$. Notice that $W'\Sigma_d^{-1/2} = O_{K'}DC_3$. With these results we can write (3.6.3) as

$$O_{K'}D^{-1}(C_3S_d^{*-1}C_3')^{-1}C_3S_d^{*-1}S_d^{*-1}C_3'(C_3S_d^{*-1}C_3')^{-1}D^{-1}O_{K'}' \quad . \tag{3.6.4}$$

Replacing $S_d^{*-1}S_d^{*-1}$ by $S_d^{*-1}(C_3'C_3 + C_4'C_4)S_d^{*-1}$ in (3.6.4) we have

$$O_{K'}D^{-2}O_{K'}' + O_{K'}D^{-1}(C_3S_d^{*-1}C_3')^{-1}C_3S_d^{*-1}C_4'C_4S_d^{*-1}C_3'(C_3S_d^{*-1}C_3')^{-1}\cdot D^{-1}O_{K'}' \; . \tag{3.6.5}$$

Since $O_{K'}D^{-2}O_{K'}' = (W'\Sigma_d^{-1}W)^{-1}$ we write the unconditional variance-covariance matrix of $\hat{\underline{\delta}}(\underline{f})$ as

$$V[\hat{\underline{\delta}}(\underline{f})] = (W'\Sigma_d^{-1}W)^{-1} + O_{K'}D^{-1}E[(C_3S_d^{*-1}C_3')^{-1}C_3S_d^{*-1}C_4'C_4S_d^{*-1}C_3' \cdot (C_3S_d^{*-1}C_3')^{-1}]D^{-1}O_{K'}' \tag{3.6.6}$$

where E stands for the expectation over the distribution of S_d^*. Although it is easy to derive the distributions of the diagonal elements of S_d^*, it is not possible to put the second term on the r.h.s. of (3.6.6) in a closed form. However, following Bement and Williams (1969) we can find some approximations to the second term. The algebraic derivation of these approximations is extremely long and quite a task. We can assert, however, that $V[\hat{\underline{\delta}}(\underline{f})]$ in (3.6.6) is finite and the second term on the r.h.s. of (3.6.6) is nonnegative definite. The first term on the r.h.s. of (3.6.6) is the variance-covariance matrix of the Aitken estimator $\hat{\underline{\delta}}(\underline{c})$. Each diagonal element of the second term indicates the magnitude of increase in the variance of the corresponding element of $\hat{\underline{\delta}}(\underline{c})$ due to replacing $\underline{c}$ by $\underline{f}$. Our intuition suggests that the diagonal elements of the second term on the r.h.s. of (3.6.6) will be small if the variances $\bar{\sigma}_1^2$, $\bar{\sigma}_2^2$ and σ_υ^2 are precisely determined.

3.7 Comparison of the Efficiencies of Pooled and OLS Estimators

The variance-covariance matrix of $\hat{\underline{\delta}}$ in (3.4.19) is

$$V(\hat{\underline{\delta}}) = (W'W)^{-1}W'\Sigma_d W(W'W)^{-1} \quad . \qquad (3.7.1)$$

Rao (1967) has shown that an Aitken estimator of a coefficient vector can be derived by making "covariance adjustment" in an OLS estimator using another statistic (with zero expectation) as a concommitant variable. Let L be a nTX(nT-K′) matrix of rank nT-K′ such that $L'W=0$. Let $\hat{\underline{\delta}}^* = L'\underline{r}$. It is easy to verify that $E\hat{\underline{\delta}}^* = 0$ and $E\hat{\underline{\delta}}^*\hat{\underline{\delta}}^{*\prime} = L'\Sigma_d L$. Covariance adjustment in $\hat{\underline{\delta}}$ is made by

$$\hat{\underline{\delta}}(\underline{c}) = \hat{\underline{\delta}} - V_{12}V_{22}^{-1}\,\hat{\underline{\delta}}^* \qquad (3.7.2)$$

where $V_{12} = \text{Cov}(\hat{\underline{\delta}},\hat{\underline{\delta}}^*) = (W'W)^{-1}W'\Sigma_d L$ and $V_{22} = V(\hat{\underline{\delta}}^*) = L'\Sigma_d L$. Consequently, we have

$$\hat{\underline{\delta}}(\underline{c}) = (W'W)^{-1}W'\underline{r} - (W'W)^{-1}W'\Sigma_d L(L'\Sigma_d L)^{-1}L'\underline{r} \qquad (3.7.3)$$

$$= (W'\Sigma_d^{-1}W)^{-1}W'\Sigma_d^{-1}\underline{r} \qquad (3.7.4)$$

where use is made of the result in Lemma 3.2.1. The variance-covariance matrix of $\hat{\underline{\delta}}(\underline{c})$ is

$$V[\hat{\underline{\delta}}(\underline{c})] = (W'\Sigma_d^{-1}W)^{-1} \quad . \qquad (3.7.5)$$

It has been shown in Lemma 3.2.2. that

$$(W'W)^{-1}W'\Sigma_d W(W'W)^{-1} - (W'\Sigma_d^{-1}W)^{-1} \qquad (3.7.6)$$

is nonnegative definite.

If we make the covariance adjustment using the estimated variances of disturbances, we obtain

$$\hat{\underline{\delta}}(\underline{f}) = \hat{\underline{\delta}} - \hat{V}_{12}\hat{V}_{22}^{-1}\hat{\underline{\delta}}^* = (W'S_d^{-1}W)^{-1}W'S_d^{-1}\underline{r} \qquad (3.7.7)$$

where $\hat{V}_{12} = (W'W)^{-1}W'S_d L$ and $\hat{V}_{22} = L'S_d L$.

Comparing (3.6.6), (3.7.1) and (3.7.6) we can say that the estimator $\hat{\underline{\delta}}(\underline{f})$ is more efficient than the estimator $\hat{\underline{\delta}}$ if each diagonal element of the matrix

(3.7.6) is larger than the corresponding diagonal element of the second term on the r.h.s. of (3.6.6). It is obvious from (3.7.3) that the Aitken estimator $\hat{\underline{\delta}}(\underline{c})$ is the same as the OLS estimator $\hat{\underline{\delta}}$ if $W'\Sigma_d L=0$. If $\hat{\underline{\delta}}(\underline{c})$ is the same as $\hat{\underline{\delta}}$, then the diagonal elements of (3.7.6) will be zero and each diagonal element of (3.7.6) cannot be larger than the corresponding diagonal element of the second term on the r.h.s. of (3.6.6). When $\sigma_\mu^2 = 0$ and $\sigma_\tau^2 = 0$, $\Sigma_d = \sigma_\upsilon^2 I_{nT}$ and $W'\Sigma_d L = 0$. Furthermore, when n, T, σ_μ^2 and σ_τ^2 are very small, the diagonal elements of Σ_d will be approximately equal to one another and $W'\Sigma_d L \approx 0$. In this case also no diagonal element of (3.7.6) may be larger than the corresponding diagonal element of the second term on the r.h.s. of (3.6.6). Consequently, when the sample sizes n and T are small, and the true values of σ_μ^2 and σ_τ^2 are small, the OLS estimator $\hat{\underline{\delta}}$ is likely to be more efficient than the Aitken estimator $\hat{\underline{\delta}}(\underline{f})$ based on $\underline{f}$.

In order to substantiate the above assertion we consider a special case where K=2 and the model (3.3.1) contains a constant term and an independent variable. In this case the variance of $\hat{\underline{\delta}}(\underline{f})$ in (3.4.16) conditional on $\underline{f}$ is

$$\left[\frac{a_1}{\hat{\sigma}_1^2} + \frac{a_2}{\hat{\sigma}_2^2} + \frac{a_3}{\hat{\sigma}_\upsilon^2}\right]^{-2} \left[\frac{\bar{\sigma}_1^2 a_1}{\hat{\sigma}_1^4} + \frac{\bar{\sigma}_2^2 a_2}{\hat{\sigma}_2^4} + \frac{\sigma_\upsilon^2 a_3}{\hat{\sigma}_\upsilon^4}\right] \tag{3.7.8}$$

where $a_1 = Z_1' N_1 Z_1$, $a_2 = Z_2' Z_2$ and $a_3 = Z_3' Z_3$. The unconditional variance of $\hat{\underline{\delta}}(\underline{f})$ is

$$E_{\underline{f}} \frac{\sigma_\upsilon^2 (h_1 c_1 + h_2 c_2 f_2^2 + f_1^2)}{a_3 (h_1 + h_2 f_2 + f_1)^2} \tag{3.7.9}$$

where $h_1 = a_1/a_3$, $h_2 = a_2/a_3$, c_1, c_2, f_1 and f_2 are as shown in (3.4.18) and (3.4.16), and E stands for the expectation over the distributions of f_1 and f_2. It is easy to show that f_1/c_1 is F-distributed with $(n-K, n'T'-K')$ d.f. and $f_2 c_2/c_1$ is also F-distributed with $(n-K, T-K)$ d.f. The variances of these F-distributions are finite if $n'T'-K' > 4$ and $T-K > 4$. For a single regressor case the variance of $\hat{\underline{\delta}}$ is

$$\frac{\sigma_\upsilon^2 (h_1 c_1 + h_2 c_2 + 1)}{a_3 (h_1 + h_2 + 1)^2} \quad . \tag{3.7.10}$$

We define the efficiency of $\hat{\underline{\delta}}(\underline{f})$ relative to that of $\hat{\underline{\delta}}$ as

$$e_1 = E_{\underline{f}} \frac{(h_1c_1 + h_2c_2f_2^2 + f_1^2)}{(h_1 + h_2f_2 + f_1)^2} \frac{(h_1 + h_2 + 1)^2}{(h_1c_1 + h_2c_2 + 1)} . \qquad (3.7.11)$$

The efficiency e_1 is a function of n, T, h_1, h_2, c_1 and c_2.

The quantities h_1 and h_2 depend upon n and T and it is difficult to know _a priori_ how they change as n and T increase. We obtain information about $\underline{\delta}$ from three sources in (3.4.2), (3.4.8) and (3.4.12). Following Theil (1963) we can interpret h_1/c_1 as the ratio of the share of information obtained from (3.4.2) to that of information obtained from (3.4.12). Similarly h_2/c_2 is the ratio of the share of information obtained from (3.4.8) to that of information obtained from (3.4.12).

We write $c_1 = 1 + T\rho_3/(1-\rho_3)$ and $c_2 = 1 + n\rho_2/(1-\rho_2)$ where $\rho_3 = \sigma_\mu^2/(\sigma_\mu^2 + \sigma_\upsilon^2)$ and ρ_2 is as shown in (3.4.9). Using a bivariate numerical integration program we evaluated e_1 for different values of T, n, h_1, h_2, ρ_2 and ρ_3. These are presented in Table 3.1. The discussion on Table 3.1 is postponed until Section 3.8.

Table 3.1

Efficiencies of the Estimators
of a Slope Coefficient

T	n	h_1	h_2	ρ_2	ρ_3	e_1	e_2	e_3
7	7	0.8	0.7	0.1	0.2	1.9756	1.4049	1.4062
				0.3	0.5	3.2324	5.2754	0.6127
				0.5	0.7	3.9385	12.8974	0.3054
				0.7	0.9	5.5907	57.5475	0.0971
10	10	0.8	0.7	0.1	0.2	2.1609	1.8247	1.1842
				0.3	0.5	3.3872	7.3163	0.4630
				0.5	0.7	3.9685	17.8848	0.2219
				0.7	0.9	5.4362	79.0066	0.0688
10	10	2.5	1.5	0.1	0.2	1.6476	0.8513	1.9353
				0.3	0.5	2.0469	2.9827	0.6863
				0.5	0.7	2.1936	6.8734	0.3191
				0.7	0.9	2.6729	28.3331	0.0943

3.8 A Comparison of the Efficiency of Pooled Estimator with those of its Components

(a) A Comparison of the Efficiency of Pooled Estimator with that of Covariance Estimator

Comparing the variance-covariance matrix of $\underset{\sim}{\tilde{\delta}}$ (which is the same as $\underline{\hat{\delta}}(3)$ in (3.4.14)) with that of $\underline{\hat{\delta}}(\underline{f})$ we can say that the Aitken estimator $\underline{\hat{\delta}}(\underline{f})$ based on $\underline{f}$ is more efficient than the covariance estimator $\underline{\hat{\delta}}(3)$ if each diagonal element of the matrix $\sigma_{\upsilon}^2(Z_3'Z_3)^{-1}-(W'\Sigma_d^{-1}W)^{-1}$ is larger than the corresponding diagonal element of the second term on the r.h.s. of (3.6.6). If the true values of σ_{μ}^2 and σ_{τ}^2 are not small and the d.f. available to estimate $\bar{\sigma}_1^2$ and $\bar{\sigma}_2^2$ (i.e., n-K and T-K) are small, then the estimated variances $\hat{\bar{\sigma}}_1^2$ and $\hat{\bar{\sigma}}_2^2$ will be large and each diagonal element of the second term on the r.h.s. of (3.6.6) will be larger than the corresponding diagonal element of the matrix $\sigma_{\upsilon}^2(Z_3'Z_3)^{-1}-(W'\Sigma_d^{-1}W)^{-1}$. In this case $\underline{\tilde{\delta}}$ will be more efficient than $\underline{\hat{\delta}}(\underline{f})$.

Now consider the special case where K=2. The efficiency of $\underline{\hat{\delta}}(\underline{f})$ relative to that of $\underline{\tilde{\delta}}$ is

$$e_2 = E_{\underline{f}} \frac{(h_1c_1 + h_2c_2f_2^2 + f_1^2)}{(h_1 + h_2f_2 + f_1)^2} \quad . \tag{3.8.1}$$

Using a bivariate numerical integration program we evaluated e_2 for different plausible values of n,T,h_1,h_2,ρ_2 and ρ_3. These values are given in Table 3.1.

The efficiency of $\underline{\tilde{\delta}}$ relative to that of $\underline{\hat{\delta}}$ is

$$e_3 = \frac{(h_1 + h_2 + 1)^2}{(h_1c_1 + h_2c_2 + 1)} \quad . \tag{3.8.2}$$

The values of e_3 for different values of n,T,h_1,h_1,ρ_2 and ρ_3 are given in Table 3.1.

From the values of e_1, e_2 and e_3 in Table 3.1 we can say that when the sample sizes n and T are around 10 and the true values of ρ_2 and ρ_3 are around 0.1 and 0.2 respectively the OLS estimator $\underline{\hat{\delta}}$ is more efficient than either the Aitken estimator $\underline{\hat{\delta}}(\underline{f})$ based on $\underline{f}$ or the covariance estimator $\underline{\tilde{\delta}}$. If the true values of ρ_2 ρ_3 are larger than 0.2 then the covariance estimator $\underline{\tilde{\delta}}$ is more efficient than either $\underline{\hat{\delta}}(\underline{f})$ or $\underline{\hat{\delta}}$. The sample sizes n and T should be sufficiently larger than 10 and the values of h_1 and h_2 should be sufficiently larger than 1 for the estimator $\hat{\underline{\delta}}(\underline{f})$ to be more efficient than either $\hat{\underline{\delta}}$ or $\tilde{\underline{\delta}}$. The values in Table 3.1 are in agreement with our assertion about the relative efficiencies of $\hat{\underline{\delta}}(\underline{f})$, $\tilde{\underline{\delta}}$ and $\hat{\underline{\delta}}$.

In order to explain what it would mean, in terms of an exogenous variable, for h_1 and h_2 to be sufficiently larger than 1 we write explicitly the forms of a_1, a_2 and a_3. If z_{1it} (i=1,2,...,n; t=1,2, ..,T) represent the observations on an independent variable, $\bar{z}_{1i\cdot} = T^{-1} \sum_{t=1}^{T} z_{1it}$, $\bar{z}_{1\cdot t} = n^{-1} \sum_{i=1}^{n} z_{1it}$ and $\bar{z}_{1\cdot\cdot} = (nT)^{-1} \cdot \sum_{i=1}^{n} \sum_{t=1}^{T} z_{1it}$, then

$$a_1 = \sum_{i=1}^{n} (\bar{z}_{1i\cdot} - \bar{z}_{1\cdot\cdot})^2, \quad a_2 = \sum_{t=1}^{T} (\bar{z}_{1\cdot t} - \bar{z}_{1\cdot\cdot})^2 \quad \text{and}$$

$$a_3 = \sum_{i=1}^{n} \sum_{t=1}^{T} (z_{1it} - \bar{z}_{1i\cdot} - \bar{z}_{1\cdot t} + \bar{z}_{1\cdot\cdot})^2 \quad . \tag{3.8.3}$$

Thus, a_1 and a_2 are the sums of squares due to variation across units and over time respectively. a_3 is the sum of squares due to remaining variation. h_1 is greater than 1 if the sum of squares due to variation across units exceeds the sum of squares due to remaining variation. Similarly, h_2 is greater than 1 if the sum of squares due to variation over time exceeds the sum of squares due to remaining variation.

(b) Comparison of the Efficiency of Pooled Estimator and with those of some of its Components

Consider the estimator

$$\hat{\underline{\delta}}(f_1) = \left[\frac{Z_1'N_1Z_1}{\hat{\sigma}_1^2} + \frac{Z_3'Z_3}{\hat{\sigma}_\upsilon^2}\right]^{-1} \left[\frac{Z_1'N_1\underline{y}_1}{\hat{\sigma}_1^2} + \frac{Z_3'\underline{y}_3}{\hat{\sigma}_\upsilon^2}\right] \tag{3.8.4}$$

which is an $\underline{\alpha}$-II estimator with $\underline{\alpha} = (1,1,0,0,\alpha_4,\alpha_5,1)'$.

If n is large, T is small and the true values of $\bar{\sigma}_1^2$ and $\bar{\sigma}_2^2$ are not small, then the d.f. available to estimate $\bar{\sigma}_2^2$ will be small. In this case for some values of parameters and for some sets of observations on the independent variables $\hat{\underline{\delta}}(f_1)$ may be more efficient than the estimators $\hat{\underline{\delta}}(\underline{f})$, $\tilde{\underline{\delta}}$, and $\hat{\underline{\delta}}$. Further investigations are necessary to prove these assertions. This problem, for the case where K=2, can be analyzed by numerical methods.

Next, consider the $\underline{\alpha}$-II estimator with $\underline{\alpha} = (0,\alpha_1,\alpha_2,1,1,0,1)'$. This is represented as

$$\hat{\underline{\delta}}(f_3) = \left[\frac{Z_2'Z_2}{\hat{\bar{\sigma}}_2^2} + \frac{Z_3'Z_3}{\hat{\sigma}_\upsilon^2}\right]^{-1}\left[\frac{Z_2'\underline{y}_2}{\hat{\bar{\sigma}}_2^2} \quad \frac{Z_3'\underline{y}_3}{\hat{\sigma}_\upsilon^2}\right] \quad . \tag{3.8.5}$$

If n is small, T is large and the true values of $\bar{\sigma}_1^2$ and $\bar{\sigma}_2^2$ are large, then for some sets of observations on the independent variables and for some values of parameters $\hat{\underline{\delta}}(f_3)$ may be more efficient than any one of the estimators $\tilde{\underline{\delta}}, \hat{\underline{\delta}}(\underline{f})$, $\hat{\underline{\delta}}(f_1)$ and $\hat{\underline{\delta}}$. We can verify this assertion, for the special case where K=2, by numerical methods.

The discussion in Sections 3.6, 3.7 and 3.8 clearly indicates that when n and/or T are small, we cannot find a single $\underline{\alpha}$-II estimator which is uniformly more efficient than any other member of $\underline{\alpha}$-II class over the entire parametric space.

3.9 Alternative Estimators of Slope Coefficients and the Regression on Lagged Values of the Dependent Variable

(a) Alternative Estimators of Slope Coefficients when the Regressors are Nonstochastic

The $\underline{\alpha}$-II class of estimators in (3.5.1) are based on $\hat{\bar{\sigma}}_1^2$, $\hat{\bar{\sigma}}_2^2$ and $\hat{\sigma}_\upsilon^2$ which are obtained by using three different estimators of the same coefficients. We ignored the restriction that the slope coefficients in (3.4.2), (3.4.8) and (3.4.12) are the same while developing the estimators of $\bar{\sigma}_1^2$, $\bar{\sigma}_2^2$ and σ_υ^2. A procedure which takes this restriction into account is Rao's (1970) principle of MINQUE. This principle involves in solving the equations given by

$$\begin{aligned} a_{11}\bar{s}_1^2 + a_{12}\bar{s}_2^2 + a_{13}s_\upsilon^2 &= \underline{w}_1'\underline{w}_1 \\ a_{21}\bar{s}_1^2 + a_{22}\bar{s}_2^2 + a_{23}s_\upsilon^2 &= \underline{w}_2'\underline{w}_2 \\ a_{31}\bar{s}_1^2 + a_{32}\bar{s}_2^2 + a_{33}s_\upsilon^2 &= \underline{w}_3'\underline{w}_3 \end{aligned} \tag{3.9.1}$$

where $\underline{w}_1 = C_2N_1\underline{y}_1 - C_2N_1Z_1\hat{\underline{\delta}}$, $\hat{\underline{\delta}}$ is as shown in (3.4.19), $\underline{w}_2 = \underline{y}_2 - Z_2\hat{\underline{\delta}}$, $\underline{w}_3 = \underline{y}_3 - Z_3\hat{\underline{\delta}}$, C_2 is as shown in (2.4.31), $a_{11} = \mathrm{tr}[I_{n'} - C_2N_1Z_1(W'W)^{-1}Z_1'N_1C_2']^2$, $a_{12} = a_{21} =$ $\mathrm{tr}\ C_2N_1Z_1(W'W)^{-1}Z_2'Z_2(W'W)^{-1}Z_1'N_1C_2'$, $a_{13} = a_{31} = \mathrm{tr}\ C_2N_1Z_1(W'W)^{-1}Z_3'Z_3(W'W)^{-1}Z_1'N_1C_2'$,

$a_{22} = \text{tr}[I_{T'} - Z_2(W'W)^{-1}Z_2']^2$, $a_{23} = a_{32} = \text{tr}\ Z_2(W'W)^{-1}Z_3'Z_3 \cdot (W'W)^{-1}Z_2'$, and $a_{33} = \text{tr}[I_{n'T'} - Z_3(W'W)^{-1}Z_3']^2$. Now suppose that $A = (a_{ij})$ is nonsingular. Then

$$\begin{bmatrix} \bar{s}_1^2 \\ \bar{s}_2^2 \\ s_\upsilon^2 \end{bmatrix} = A^{-1} \begin{bmatrix} \underline{w}_1'\underline{w}_1 \\ \underline{w}_2'\underline{w}_2 \\ \underline{w}_3'\underline{w}_3 \end{bmatrix} . \qquad (3.9.2)$$

The variances $\bar{\sigma}_1^2$, $\bar{\sigma}_2^2$ and σ_υ^2 in the Aitken estimator $\hat{\underline{\delta}}(\underline{c})$ can be replaced by their estimates in (3.9.2) to obtain a feasible estimator for $\underline{\delta}$. This estimator may be unbiased and more efficient than any member of $\underline{\alpha}$-II class estimators for some values of the parameters in small samples. Analytical study of this problem is difficult. Monte Carlo studies may throw some light.

(b) Regression on the Lagged Values of the Dependent Variable

Suppose that the specification (1) in Assumption 3.5.1 is not true and the lagged values of the dependent variable appear as explanatory variables. Even in this case if τ_t and υ_{it} are serially independent, the asymptotic results in Section 3.5 are valid because under very general conditions stated in Goldberger (1964, pp. 272-4)

$$\underset{\substack{n\to\infty\\T\to\infty}}{\text{plim}}(nT)^{-1}X_1'\underline{u}_1 = \underset{\substack{n\to\infty\\T\to\infty}}{\text{plim}}(nT)^{-1}X'(I_n \otimes \underline{\iota}_T/\sqrt{T})(I_n \otimes \underline{\iota}_T'/\sqrt{T})\underline{u}=0,$$

$$\underset{\substack{n\to\infty\\T\to\infty}}{\text{plim}}(nT)^{-1}Z_2'\underline{u}_2 = \underset{\substack{n\to\infty\\T\to\infty}}{\text{plim}}(nT)^{-1}Z'(\underline{\iota}_n/\sqrt{n} \otimes C_1')(\underline{\iota}_n'/\sqrt{n} \otimes C_1)\underline{u}=0, \text{ and}$$

$$\underset{\substack{n\to\infty\\T\to\infty}}{\text{plim}}(nT)^{-1}Z_3'\underline{u}_3 = \underset{\substack{n\to\infty\\T\to\infty}}{\text{plim}}(nT)^{-1}Z'(C_2' \otimes C_1')(C_2 \otimes C_1)\underline{u}=0 .$$

The specification (2) in Assumption 3.5.1 is tenable with a reinterpretation of "lim" as "plim."

$\hat{\underline{\delta}}(\underline{f})$ is a consistent estimator of $\underline{\delta}$ and we can apply it to estimate the coefficients in a dynamic equation of the type (3.3.1) provided the τ_t and υ_{it} are serially independent. The small sample results in Sections 3.6 and 3.7 are not valid. The estimates in (3.9.2) are inconsistent because they are based on the residuals obtained by OLS, applied directly to (3.3.1).

3.10 Analysis of An Error Components Model Under Alternative Assumptions

Consider the model (3.3.1) and the following assumption.

Assumption 3.10.1:

(1) $T > K$ and $n > 1$.

(2) Same as (2) in Assumption 2.4.1.

(3) The μ_i's are fixed parameters obeying the restriction $\sum_{i=1}^{n} \mu_i = 0$.

(4) Same as (4) in Assumption 3.4.1.

(5) Same as (4) in Assumption 2.4.1.

(6) τ_t and υ_{it} are independent of each other.

This assumption was adopted by Hussain (1969).

The variance-covariance matrix of the error term in (3.3.1) under Assumption 3.10.1 is

$$E\underline{u}\underline{u}' = \sigma_\tau^2 (\underline{\iota}_n \underline{\iota}_n' \otimes I_T) + \sigma_\upsilon^2 I_{nT} \quad .$$

Making use of the transformations defined by Q_2 and Q_3 in (3.4.8) and (3.4.12) we obtain an estimator for $\underline{\delta}$ as

$$\hat{\underline{\delta}}(f_3) = \left[\frac{Z_2'Z_2}{\hat{\sigma}_2^2} + \frac{Z_3'Z_3}{\hat{\sigma}_\upsilon^2}\right]^{-1} \left[\frac{Z_2'\underline{y}_2}{\hat{\sigma}_2^2} + \frac{Z_3'\underline{y}_3}{\hat{\sigma}_\upsilon^2}\right] \quad . \qquad (3.10.2)$$

where $f_3 = \hat{\sigma}_2^2/\hat{\sigma}_\upsilon^2$. Notice that (3.10.2) is the same as (3.8.5).

Following the analytical and numerical methods discussed in Subsection 2.4(c) we can study the small sample properties of $\hat{\underline{\delta}}(f_3)$.

Instead of Assumption 3.10.1 we could as well consider the following assumption.

Assumption 3.10.2:

(1) $T > 1$ and $n > K$.

(2) Same as (2) in Assumption 2.4.1.

(3) Same as (3) in Assumption 2.4.1.

(4) The τ_t's are fixed parameters obeying the restriction $\sum_{t=1}^{T} \tau_t = 0$.

(5) Same as (4) in Assumption 2.4.1.

(6) μ_i and υ_{it} are independent of each other.

Under this assumption (3.8.4) is an Aitken estimator of $\underline{\delta}$ based on estimated variances.

3.11 Maximum Likelihood Method of Estimating Error Components Model

Under Assumptions 3.4.1 and 3.5.2 we can write the likelihood function of parameters in (3.3.1) as

$$L(\underline{\beta},\underline{c}|\underline{y},X) = (2\pi)^{-nT/2}(\bar{\sigma}_1^2 + \bar{\sigma}_2^2 - \sigma_\upsilon^2)^{-1/2}(\bar{\sigma}_1^2)^{-n'/2}(\bar{\sigma}_2^2)^{-T'/2}$$

$$\cdot(\sigma_\upsilon^2)^{-n'T'/2} \exp\left\{-\frac{1}{2}\left[\frac{(\iota_n'\underline{y}_1 - \iota_n'X_1\underline{\beta})^2}{n(\bar{\sigma}_1^2 + \bar{\sigma}_2^2 - \sigma_\upsilon^2)} + \frac{(\underline{y}_1 - Z_1\underline{\delta})'N_1(\underline{y}_1 - Z_1\underline{\delta})}{\bar{\sigma}_1^2} + \frac{(\underline{y}_2 - Z_2\underline{\delta})'(\underline{y}_2 - Z_2\underline{\delta})}{\bar{\sigma}_2^2} + \frac{(\underline{y}_3 - Z_3\underline{\delta})'(\underline{y}_3 - Z_3\underline{\delta})}{\sigma_\upsilon^2}\right]\right\}, \qquad (3.11.1)$$

where the symbols are as explained in Section 3.4.

If we differentiate $\log L(\underline{\beta},\underline{c}|\underline{y},X)$ with respect to $\beta_0, \underline{\delta}, \bar{\sigma}_1^2, \bar{\sigma}_2^2$, and σ_υ^2 and equate them to zero, we obtain a set of normal equations which are nonlinear in the unknowns. However, the reparameterization of the model in terms of ρ_2 and ρ_3 suggests a simple procedure of solving these equations.

We can write (3.11.1) as

$$L(\underline{\beta}, \rho_2, \rho_3, \sigma_\upsilon^2 | \underline{y}, X) = (2\pi)^{-nT/2} (\sigma_\upsilon^2)^{-nT/2} \left\{T\rho_3(1-\rho_3)^{-1} + n\rho_2(1-\rho_2)^{-1} + 1\right\}^{-1/2}$$

$$\cdot \left\{1 + T\rho_3(1-\rho_3)^{-1}\right\}^{-n'/2} \left\{1 + n\rho_2(1-\rho_2)^{-1}\right\}^{-T'/2}$$

$$\cdot \exp\left\{-\frac{1}{2\sigma_\upsilon^2}\left[(\underline{\iota}_n'\underline{y}_1 - \underline{\iota}_n'X_1\underline{\beta})^2 \, n^{-1} \left(\frac{T\rho_3}{1-\rho_3} + \frac{n\rho_2}{1-\rho_2} + 1\right)^{-1}\right.\right.$$

$$\left.\left. + (\underline{r} - W\underline{\delta})' \, \Sigma_d^{*-1} \, (\underline{r} - W\underline{\delta})\right]\right\} \quad , \qquad (3.11.2)$$

where $\Sigma_d^* = \text{diag}\left[\left(1 + \frac{T\rho_3}{1-\rho_3}\right) I_n, \left(1 + \frac{n\rho_2}{1-\rho_2}\right) I_{T'}, I_{n'T'}\right]$.

If we differentiate the logarithmic likelihood function with respect to β_0 we obtain

$$\frac{\partial \log L}{\partial \beta_0} = \frac{\sqrt{T}\,\underline{\iota}_n'\underline{y}_1 - nT\beta_0 - \underline{\delta}'Z_1'\underline{\iota}_n\sqrt{T}}{\sigma_\upsilon^2\left(\frac{T\rho_3}{1-\rho_3} + \frac{n\rho_2}{1-\rho_2} + 1\right)} \quad . \qquad (3.11.3)$$

Equating (3.11.3) to zero and solving for β_0 we have

$$\check{\beta}_0(\underline{\delta}) = \frac{1}{n\sqrt{T}} (\underline{\iota}_n'\underline{y}_1 - \underline{\iota}_n'Z_1\underline{\delta}) \quad . \qquad (3.11.4)$$

Similarly, if we differentiate the logarithmic likelihood function with respect to $\underline{\delta}$ we obtain

$$\frac{\partial \log L}{\partial \underline{\delta}} = \frac{Z_1'\underline{\iota}_n\underline{\iota}_n'\underline{y}_1 - Z_1'\underline{\iota}_n\beta_0 n\sqrt{T} - Z_1'\underline{\iota}_n\underline{\iota}_n'Z_1\underline{\delta}}{n\,\sigma_\upsilon^2\left(\frac{n\rho_2}{1-\rho_2} + \frac{T\rho_3}{1-\rho_3} + 1\right)} + \frac{W'\Sigma_d^{*-1}\underline{r} - W'\Sigma_d^{*-1}W\underline{\delta}}{\sigma_\upsilon^2} \quad . \qquad (3.11.5)$$

Equating (3.11.5) to zero and substituting (3.11.4) into (3.11.5), we have

$$\check{\underline{\delta}}(\rho_2, \rho_3) = (W'\Sigma_d^{*-1}W)^{-1}W'\Sigma_d^{*-1}\,\underline{r} \quad . \qquad (3.11.6)$$

Differentiating log L with respect to σ_υ^2 we have

$$\frac{\partial \log L}{\partial \sigma_\upsilon^2} = -\frac{nT}{2}\frac{1}{\sigma_\upsilon^2} + \frac{1}{2\sigma_\upsilon^4}\Big[(\underline{\iota}_n'\underline{y}_1 - \underline{\iota}_n'X_1\underline{\beta})^2 n^{-1}\left(\frac{T\rho_3}{1-\rho_3} + \frac{n\rho_2}{1-\rho_2} + 1\right)^{-1}$$

$$+ (\underline{r} - W\underline{\delta})' E_d^{*-1} (\underline{r} - W\underline{\delta})\Big] \quad . \tag{3.11.7}$$

Equating (3.11.7) to zero and substituting (3.11.4) and (3.11.6) into (3.11.7), we have

$$\check{\sigma}_\upsilon^2 (\rho_2,\rho_3) = \frac{[\underline{r} - W\check{\underline{\delta}}(\rho_2,\rho_3)]'\Sigma_d^{*-1}[\underline{r} - W\check{\underline{\delta}}(\rho_2,\rho_3)]}{nT} \quad . \tag{3.11.8}$$

Inserting (3.11.4), (3.11.6) and (3.11.8) in the likelihood function, we obtain a concentrated likelihood function, say $\hat{L}$, which we may write purely as a function of ρ_2,ρ_3 and observations.

$$\log \hat{L}(\rho_2,\rho_3) = -\frac{nT}{2}(1 + \log 2\pi) - \frac{nT}{2}\log \check{\sigma}_\upsilon^2 (\rho_2,\rho_3)$$

$$- \frac{1}{2}\log\left\{T\rho_3(1-\rho_3)^{-1} + n\rho_2(1-\rho_2)^{-1} + 1\right\}$$

$$- \frac{n'}{2}\log\left\{1 + T\rho_3(1-\rho_3)^{-1}\right\} - \frac{T'}{2}$$

$$\cdot \log\left\{1 + n\rho_2(1-\rho_2)^{-1}\right\} \quad . \tag{3.11.9}$$

For different values of (ρ_2,ρ_3) in the region $(0 \le \rho_2 < 1;\ 0 \le \rho_3 < 1)$, we can obtain the values of $\check{\underline{\delta}}(\rho_2,\rho_3)$, $\check{\sigma}_\upsilon^2(\rho_2,\rho_3)$ and $\hat{L}(\rho_2,\rho_3)$. Since $\hat{L}(\rho_2,\rho_3)$ is continuous in the region $(0 \le \rho_2 < 1;\ 0 \le \rho_3 < 1)$, it may reach a maximum within the region. If $\hat{L}(\rho_2,\rho_3)$ reaches a maximum within the region, we can find the value of (ρ_2,ρ_3) for which this occurs by computing $\hat{L}(\rho_2,\rho_3)$ numerically for a sufficient number of points (ρ_2,ρ_3) within the region $(0 \le \rho_2 < 1;\ 0 \le \rho_3 < 1)$. The values of β_0, $\underline{\delta}$, ρ_2, ρ_3, and σ_υ^2 corresponding to sup $\hat{L}(\rho_2,\rho_3)$ are the ML estimates of β_0, $\underline{\delta}$, ρ_2, ρ_3, and σ_υ^2, respectively. It can be shown that these ML estimates are consistent and asymptotically normal.

A similar iterative maximum likelihood procedure was adopted by Nerlove (1965, pp. 164-173) to estimate the parameters of (1.2.1) with the error structure (1.2.2). Notice that the structure (1.2.2) is the same as the error structure of (3.3.1) under Assumption 3.4.1 when $\sigma_\tau^2 = 0$. The discussion in this chapter clearly indicates that we can obtain consistent estimates of the parameters in (1.2.1) even when $\sigma_\tau^2 \neq 0$. It is not clear why Nerlove (1965, p. 160) says that the inclusion of a separate time effect, τ_t, would greatly complicate the analysis without adding any essential generality.

3.12 Departures from the Basic Assumptions Underlying the Error Components Model

(a) Robustness of Inferences Based on the Pooled Estimator

In this section we will show that the Aitken procedure based on estimated error variances leads to the estimator $\hat{\underline{\delta}}(\underline{f})$ under the following assumption which is more general than Assumption 3.4.1.

Assumption 3.12.1:

(1) Same as (1) in Assumption 2.6.1.

(2) Same as (2) in Assumption 3.3.1.

(3) For i, i′ = 1,2,...,n:

$$E\mu_i = 0;$$

$$E\mu_i\mu_{i'} = \begin{cases} \sigma_\mu^2 & \text{if } i = i', \\ \rho_4\sigma_\mu^2 & \text{if } i \neq i'; \end{cases}$$

$$-(n-1)^{-1} \le \rho_4 \le 1 .$$

(4) For t, t′ = 1,2,...,T:

$$E\tau_t = 0;$$

$$E\tau_t\tau_{t'} = \begin{cases} \sigma_\tau^2 & \text{if } t = t', \\ \rho_5\sigma_\tau^2 & \text{if } t \neq t'; \end{cases}$$

$$-(T-1)^{-1} \le \rho_5 \le 1 .$$

(5) For $i,i' = 1,2,\ldots,n$ and $t,t' = 1,2,\ldots,T$:

$$E\upsilon_{it} = 0 \; ;$$

$$E\upsilon_{it}\upsilon_{i't'} = \begin{cases} \sigma_\upsilon^2 \text{ if } i = i' \text{ and } t = t', \\ \rho_6\sigma_\upsilon^2 \text{ if } i \neq i' \text{ and } t = t', \\ \rho_7\sigma_\upsilon^2 \text{ if } i = i' \text{ and } t \neq t', \\ \rho_6\rho_7\sigma_\upsilon^2 \text{ if } i \neq i' \text{ and } t \neq t'; \end{cases}$$

$$-(n-1)^{-1} < \rho_6 < 1 \text{ and } -(T-1)^{-1} < \rho_7 < 1.$$

(6) μ_i, τ_t and υ_{it} are mutually uncorrelated.

Assumption 3.4.1 is a special case of Assumption 3.12.1. If $\rho_i = 0 (i = 4,5,6,7)$ and μ_i and $\underline{\upsilon}_i$ are normal, then Assumption 3.12.1 is the same as Assumption 3.4.1. As has been pointed out by Nerlove (1965, p. 160) the most dubious specification in Assumption 3.4.1 is that the μ_i and υ_{it} are independent of themselves for different individuals. For example, if the individuals were geographical regions with arbitrarily drawn boundaries, we would hardly expect this assumption to be satisfied in practice. When we are analyzing regionwise data, it may be desirable to assume that the $\mu_i (i = 1,2,\ldots,n)$ and the $\upsilon_{it} (i = 1,2,\ldots,n)$ are pairwise equicorrelated as in specifications (3) and (5) of Assumption 3.12.1. The specifications (4), (5), and (6) in Assumption 3.12.1 are also weaker than the specifications (4), (5), and (6), respectively in Assumption 3.4.1.

Under Assumption 3.12.1

$$E\underline{u}\underline{u}' = \sigma_\mu^2 (\Omega_4 \otimes \underline{\iota}_T\underline{\iota}_T') + \sigma_\tau^2 (\underline{\iota}_n\underline{\iota}_n' \otimes \Omega_5) + \sigma_\upsilon^2 (\Omega_6 \otimes \Omega_7) \tag{3.12.1}$$

where

$$\Omega_4 = [\rho_4\underline{\iota}_n\underline{\iota}_n' + (1-\rho_4)I_n], \; \Omega_5 = [\rho_5\underline{\iota}_T\underline{\iota}_T' + (1-\rho_5)I_T] \; ,$$

$$\Omega_6 = [\rho_6\underline{\iota}_n\underline{\iota}_n' + (1-\rho_6)I_n], \text{ and } \Omega_7 = [\rho_7\underline{\iota}_T\underline{\iota}_T' + (1-\rho_7)I_T], \tag{3.12.2}$$

$$E\underline{u}_1 = 0 \quad , \tag{3.12.3}$$

$$E\underline{u}_1\underline{u}_1' = (I_n \otimes \underline{\iota}_T'/\sqrt{T})[\sigma_\mu^2 (\Omega_4 \otimes \underline{\iota}_T\underline{\iota}_T') + \sigma_\tau^2(\underline{\iota}_n\underline{\iota}_n' \otimes \Omega_5)$$

$$+ \sigma_\upsilon^2 (\Omega_6 \otimes \Omega_7)](I_n \otimes \underline{\iota}_T/\sqrt{T})$$

$$= [T\ \sigma_\mu^2\rho_4 + \sigma_\tau^2 (1 - \rho_5 + T\rho_5) + \sigma_\upsilon^2\rho_6 (1 - \rho_7 + T\rho_7)]\underline{\iota}_n\underline{\iota}_n'$$

$$+ [T\ \sigma_\mu^2(1-\rho_4) + \sigma_\upsilon^2 (1 - \rho_7 + T\rho_7)(1-\rho_6)]I_n \quad , \qquad (3.12.4)$$

$$E\underline{u}_2 = 0 \quad , \qquad (3.12.5)$$

$$E\underline{u}_2\underline{u}_2' = (\underline{\iota}_n'/\sqrt{n} \otimes C_1)[\sigma_\mu^2(\Omega_4 \otimes \underline{\iota}_T\underline{\iota}_T') + \sigma_\tau^2 (\underline{\iota}_n\underline{\iota}_n' \otimes \Omega_5)$$

$$+ \sigma_\upsilon^2 (\Omega_6 \otimes \Omega_7)](\underline{\iota}_n/\sqrt{n} \otimes C_1')$$

$$= [n\ \sigma_\tau^2 (1-\rho_5) + \sigma_\upsilon^2 (1-\rho_7)(1-\rho_6 + n\rho_6)I_{T'} \quad , \qquad (3.12.6)$$

$$E\underline{u}_3 = 0, \text{ and} \qquad (3.12.7)$$

$$E\underline{u}_3\underline{u}_3' = (C_2 \otimes C_1)[\sigma_\mu^2 (\Omega_4 \otimes \underline{\iota}_T\underline{\iota}_T') + \sigma_\tau^2 (\underline{\iota}_n\underline{\iota}_n' \otimes \Omega_5)$$

$$+ \sigma_\upsilon^2 (\Omega_6 \otimes \Omega_7)](C_2' \otimes C_1')$$

$$= \sigma_\upsilon^2 (1-\rho_6)(1-\rho_7)I_{n'T'} \quad . \qquad (3.12.8)$$

Consequently, the estimator (3.4.4) is the BLUE of $\underline{\beta}$ in the setup (3.4.2), the estimator (3.4.10) is the BLUE of $\underline{\delta}$ in the setup (3.4.8) and the estimator (3.4.14) is the BLUE of $\underline{\delta}$ in the setup (3.4.12). The estimators (3.4.7), (3.4.11), and (3.4.15) are the unbiased estimators of $[T\ \sigma_\mu^2(1-\rho_4) + \sigma_\upsilon^2(1-\rho_7 + T\rho_7)(1-\rho_6)]$, $[n\ \sigma_\tau^2(1-\rho_5) + \sigma_\upsilon^2(1-\rho_7)(1-\rho_6 + n\rho_6)]$ and $\sigma_\upsilon^2(1-\rho_6)(1-\rho_7)$, respectively. The estimator $\hat{\underline{\delta}}(\underline{f})$ is still an Aitken estimator of $\underline{\delta}$ based on estimated variances of errors. The properties of $\hat{\underline{\delta}}(\underline{f})$ derived in Section 3.6 remain unchanged even when Assumption 3.12.1 is true, if we reinterpret the parameters $\overline{\sigma}_1^2$ $\overline{\sigma}_2^2$, and σ_υ^2 as $[T\ \sigma_\mu^2(1-\rho_4) + \sigma_\upsilon^2(1-\rho_7 + T\rho_7)(1-\rho_6)]$, $[n\ \sigma_\tau^2(1-\rho_5) + \sigma_\upsilon^2(1-\rho_7)(1-\rho_6 + n\rho_6)]$ and $\sigma_\upsilon^2(1-\rho_6)(1-\rho_7)$, respectively. Notice that under this interpretation the restrictions $\overline{\sigma}_1^2 \geq \sigma_\upsilon^2$ and $\overline{\sigma}_2^2 \geq \sigma_\upsilon^2$ are no longer present. There is no weakness in the Aitken procedure when there are no restrictions on the variances of $\underline{u}_1$, $\underline{u}_2$, and $\underline{u}_3$.

(b) Sensitivity of ML Estimates of Coefficients in the Error Components Model

The ML procedure developed in Section 3.11 takes into account the restrictions $\bar{\sigma}_1^2 \geq \sigma_\upsilon^2$ and $\bar{\sigma}_2^2 \geq \sigma_\upsilon^2$. If Assumption 3.4.1 is correct, then these restrictions are present and the ML estimates under these restrictions can be found if the function (3.11.9) attains a maximum in the region $(0 \leq \rho_2 < 1, 0 \leq \rho_3 < 1)$. Nerlove (1965), p. 173) considered the likelihood function (3.11.2) under the restriction $\rho_2 = 0$, and remarked that there was no guarantee, for any particular $\underline{y}$ and X, that the maximum of log $\hat{L}(\rho_3)$ actually would be within the region $0 \leq \rho_3 < 1$. Indeed, in an analysis of a time series of cross-sections related to natural gas consumption, Balestra and Nerlove (1966) found that log $\hat{L}(\rho_3)$ reached a maximum for $\rho_3 = 0$ and even went on increasing for negative values of ρ_3. This led Balestra and Nerlove to discard the ML approach. Our analysis in subsection 3.12(a) partly explains their result.

Now consider the model (3.3.1). Let the matrix Z contain some lagged values of the dependent variable. We utilize ML method to estimate $\underline{\beta}$ in (3.3.1) when Assumption 3.12.1 holds. We suppose, as is usual in such circumstances, that the lagged values of the dependent variable are fixed. Under Assumptions 3.12.1 and 3.5.2 the likelihood function of parameters is given by

$$L(\underline{\beta},\underline{\sigma}|\underline{y},X) = (2\pi)^{-nT/2} |\Omega_8|^{-1/2} (\sigma_5^{\,2})^{-T'/2} (\sigma_6^{\,2})^{-n'T'/2}$$

$$\exp\left\{-\frac{1}{2}\left[(\underline{y}_1 - X_1\underline{\beta})'\Omega_8^{-1}(\underline{y}_1 - X_1\underline{\beta}) + \frac{(\underline{y}_2 - Z_2\underline{\delta})'(\underline{y}_2 - Z_2\underline{\delta})}{\sigma_5^{\,2}} + \frac{(\underline{y}_3 - Z_3\underline{\delta})'(\underline{y}_3 - Z_3\underline{\delta})}{\sigma_6^{\,2}}\right]\right\} \tag{3.12.9}$$

where $\Omega_8 = E\underline{u}_1\underline{u}_1'$ in (3.12.4), $\sigma_5^{\,2} = [n\sigma_\tau^{\,2}(1-\rho_5) + \sigma_\upsilon^{\,2}(1-\rho_7)(1-\rho_6+n\rho_6)]$, $\sigma_6^{\,2} = \sigma_\upsilon^{\,2}(1-\rho_6)(1-\rho_7)$.

Using the result (i) in Lemma 2.2.1 we can show that

$$|\Omega_8| = (\sigma_4^{\,2})^{n-1}\,\sigma_3^{\,2} \tag{3.12.10}$$

where $\sigma_4^{\,2} = [T\,\sigma_\mu^{\,2}(1-\rho_4) + \sigma_\upsilon^{\,2}(1-\rho_7+T\rho_7)(1-\rho_6)]$ and $\sigma_3^{\,2} = \sigma_4^{\,2} + n[T\,\sigma_\mu^{\,2}\rho_4 + \sigma_\tau^{\,2}\cdot(1-\rho_5+T\rho_5) + \sigma_\upsilon^{\,2}\rho_6(1-\rho_7+T\rho_7)]$.

It follows from Lemma 2.2.2 that

$$\Omega_8^{-1} = (\sigma_4^{\,2})^{-1}I_n - (\sigma_4^{\,2})^{-1}(\sigma_3^{\,2})^{-1}\,[T\,\sigma_\mu^{\,2}\rho_4 + \sigma_\tau^{\,2}(1 - \rho_5 + T\rho_5)$$

$$+ \sigma_\upsilon^{\,2}\rho_6(1 - \rho_7 + T\rho_7)]\underline{\iota}_n\underline{\iota}_n' - \underline{\iota}_n\underline{\iota}_n'(1/n\sigma_4^{\,2}) + \underline{\iota}_n\underline{\iota}_n'(1/n\sigma_4^{\,2})$$

$$= (\sigma_4^{\,2})^{-1}\,N_1 + \underline{\iota}_n\underline{\iota}_n'(1/n\sigma_3^{\,2}) \quad . \tag{3.12.11}$$

We substitute (3.12.10) and 3.12.11) in (3.12.9) to obtain

$$L(\underline{\beta},\underline{\sigma}|\underline{y},X) = (2\pi)^{-nT/2}(\sigma_3^{\,2})^{-1/2}(\sigma_4^{\,2})^{-n'/2}(\sigma_5^{\,2})^{-T'/2}(\sigma_6^{\,2})^{-n'T'/2}$$

$$\cdot \exp\left\{-\frac{1}{2}\left[\frac{(\underline{\iota}_n'\underline{y}_1 - \underline{\iota}_n'X_1\underline{\beta})^2}{n\sigma_3^{\,2}} + \frac{(\underline{y}_1 - Z_1\underline{\delta})'N_1(\underline{y}_1 - Z_1\underline{\delta})}{\sigma_4^{\,2}}\right.\right.$$

$$+ \frac{(\underline{y}_2 - Z_2\underline{\delta})'\,(\underline{y}_2 - Z_2\underline{\delta})}{\sigma_5^{\,2}}$$

$$\left.\left. + \frac{(\underline{y}_3 - Z_3\underline{\delta})'\,(\underline{y}_3 - Z_3\underline{\delta})}{\sigma_6^{\,2}}\right]\right\} \quad . \tag{3.12.12}$$

The likelihood functions (3.11.1) and (3.12.12) are indistinguishable. The restriction $\sigma_4^{\,2} \geq \sigma_6^{\,2}$ is present either if $\rho_7 \geq 0$ or if $\rho_7 < 0$ and $\sigma_\mu^{\,2}(1-\rho_4) \geq \sigma_\upsilon^{\,2}(1-\rho_6)|\rho_7|$. Similarly, the restriction $\sigma_5^{\,2} \geq \sigma_6^{\,2}$ is present as long as $\rho_6 \geq 0$ or $\rho_6 < 0$ and $\sigma_\tau^{\,2}(1-\rho_5) \geq \sigma_\upsilon^{\,2}|\rho_6|(1-\rho_7)$. The variance function $\sigma_3^{\,2}$ is not estimable. The variances $\sigma_\mu^{\,2}$, $\sigma_\tau^{\,2}$ and $\sigma_\upsilon^{\,2}$ are not estimable unless $\rho_4 = \rho_5 = \rho_6 = \rho_7 = 0$. We can write

$$L(\underline{\beta},\underline{\sigma}|\underline{y},X) = (2\pi)^{-nT/2}(\sigma_6^{\,2})^{-nT/2}\left\{1 + T\bar{\rho}_3(1-\bar{\rho}_3)^{-1} + n\bar{\rho}_2(1-\bar{\rho}_2)^{-1} + nT\bar{\rho}_0\right\}^{-1/2}$$

$$\cdot\left\{1 + T\bar{\rho}_3(1-\bar{\rho}_3)^{-1}\right\}^{-n'/2}\left\{1 + n\bar{\rho}_2(1-\bar{\rho}_2)^{-1}\right\}^{-T'/2}$$

$$\cdot \exp\left\{-\frac{1}{2\sigma_6^2}\left[(\underline{\iota}_n'\underline{y}_1 - \underline{\iota}_n'X_1\underline{\beta})^2\left(\frac{1}{n}\right)\left(1+\frac{T\bar{\rho}_3}{(1-\bar{\rho}_3)}+\frac{n\bar{\rho}_2}{(1-\bar{\rho}_2)}+nT\bar{\rho}_0\right)^{-1}\right.\right.$$

$$\left.\left. + (\underline{r} - W\underline{\delta})'\,\bar{\Sigma}_d^{*-1}\,(\underline{r} - W\underline{\delta})\right]\right\} \qquad (3.12.13)$$

where

$$\bar{\rho}_0 = (\sigma_\mu^2\rho_4 + \sigma_\tau^2\rho_5 + \sigma_\upsilon^2\rho_6\rho_7)/\sigma_6^2 \quad ,$$

$$\bar{\Sigma}_d^* = \text{diag}\left[\left(1+\frac{T\bar{\rho}_3}{1-\bar{\rho}_3}\right)I_n,\ \left(1+\frac{n\bar{\rho}_2}{1-\bar{\rho}_2}\right)I_{T'},\ I_{n'T'}\right] \quad ,$$

$$\bar{\rho}_2 = [\sigma_\tau^2(1-\rho_5) + \sigma_\upsilon^2(1-\rho_7)\rho_6]/[\sigma_\tau^2(1-\rho_5) + \sigma_\upsilon^2(1-\rho_7)] \quad \text{and}$$

$$\bar{\rho}_3 = [\sigma_\mu^2(1-\rho_4) + \sigma_\upsilon^2(1-\rho_6)\rho_7]/[\sigma_\mu^2(1-\rho_4) + \sigma_\upsilon^2(1-\rho_6)] \ .$$

In this case $\bar{\rho}_2$ and $\bar{\rho}_3$ can be negative and the likelihood function (3.12.13) need not necessarily attain its maximum in the region $(0 \le \bar{\rho}_2 < 1;$ $0 \le \bar{\rho}_3 < 1)$. Because of this possibility if the maximum of log $\hat{L}(\rho_2,\rho_3)$ in (3.11.9) does not lie within the region $(0 \le \rho_2 < 1;\ 0 \le \rho_3 < 1)$, we may conclude that the specifications (3), (4), (5) and (6) in Assumption 3.4.1 are inappropriate. If the specifications (3), (4), (5), and (6) in Assumption 3.4.1 are invalid, we may obtain implausible estimates through the ML procedure outlined in Section 3.11, and we may obtain negative estimates for σ_μ^2 and σ_τ^2 if we take $\hat{\bar{\sigma}}_1^2$, $\hat{\bar{\sigma}}_2^2$, and $\hat{\sigma}_\upsilon^2$ as the estimators of $(T\,\sigma_\mu^2 + \sigma_\upsilon^2)$, $(n\,\sigma_\tau^2 + \sigma_\upsilon^2)$ and σ_υ^2, respectively.

3.13 Conclusions

We considered, in this chapter, the use of error components models in the analysis of combined time series and cross-section data. We developed an Aitken estimator of a coefficient vector based on estimated variance components and studied its finite sample properties. Out study indicates that this estimator is less efficient than an OLS estimator if the sample sizes n and T, and the true values of σ_μ^2 and σ_τ^2 are small. It is also less efficient than a covariance estimator if the sample sizes n and T are small and the true values of σ_μ^2 and σ_τ^2 are not small. Finally, we have developed an $\underline{\alpha}$-II class of estimators and shown that it is not possible to choose an estimator of $\underline{\delta}$ on the basis of asymptotic efficiency without bringing in small sample considerations.

Following Rao (1970) we found minimum norm quadratic unbiased estimators of σ_1^2, σ_2^2, and σ_υ^2 and substituted them in the Aitken estimator $\hat{\underline{\delta}}(\underline{c})$. Such an estimator of $\underline{\delta}$ may have some desirable small sample properties.

The estimator $\hat{\underline{\delta}}(\underline{f})$ is consistent under very general conditions even if the lagged values of the dependent variable appear as explanatory variables provided τ_t and υ_{it} are serially independent. An iterative ML method of estimating the parameters in error components models is developed in the last but one section of this chapter. This method yields consistent estimates.

The estimator $\hat{\underline{\delta}}(\underline{f})$ is an Aitken estimator based on estimated variances of errors even when the elements of each error component are pairwise equicorrelated.

APPENDIX TO CHAPTER III

A. Confidence Intervals for the Ratios of Variance Functions

We have shown in subsection 3.12(a) that under Assumption 3.12.1 $\hat{\bar{\sigma}}_1^2$ in (3.4.7) is an unbiased estimator of the variance function $\sigma_4^2 = [T\sigma_\mu^2(1-\rho_4) + \sigma_\nu^2 \cdot(1-\rho_7+T\rho_7)(1-\rho_6)]$, $\hat{\bar{\sigma}}_2^2$ in (3.4.11) is an unbiased estimator of the variance function $\sigma_5^2 = [n\sigma_\tau^2(1-\rho_5) + \sigma_\nu^2(1-\rho_7)(1-\rho_6+n\rho_6)]$, $\hat{\sigma}_\nu^2$ in (3.4.15) is an unbiased estimator of the variance function $\sigma_6^2 = \sigma_\nu^2(1-\rho_6)(1-\rho_7)$. Under Assumption 3.5.2 the functions $\hat{\bar{\sigma}}_1^2$, $\hat{\bar{\sigma}}_2^2$ and $\hat{\sigma}_\nu^2$ are independent of each other because $\underline{y}_1$, $\underline{y}_2$ and $\underline{y}_3$ are mutually independent. Since M_1, M_2 and M_3 are idempotent matrices, $(n-K)\,\hat{\bar{\sigma}}_1^2/\sigma_4^2$, $(T-K)\,\hat{\bar{\sigma}}_2^2/\sigma_5^2$ and $[(n-1)(T-1)-K+1]\,\hat{\sigma}_\nu^2/\sigma_6^2$ are distributed as χ^2 with (n-K), (T-K) and [(n-1)(T-1)-K+1] d.f. respectively. Consequently, $\hat{\bar{\sigma}}_1^2\sigma_6^2/\hat{\sigma}_\nu^2\sigma_4^2$ and $\hat{\bar{\sigma}}_2^2\sigma_6^2/\hat{\sigma}_\nu^2\sigma_5^2$ are distributed as F with (n-K), [(n-1)(T-1)-K+1] and (T-K), [(n-1)(T-1)-K+1] d.f. respectively. Given $\alpha > 0$, we can find four values $F_1(\alpha)$, $F_2(\alpha)$, $F_3(\alpha)$ and $F_4(\alpha)$ such that

$$\Pr\left[F_1(\alpha) \le \frac{\hat{\bar{\sigma}}_1^2\sigma_6^2}{\hat{\sigma}_\nu^2\sigma_4^2} \le F_2(\alpha)\right] = 1-\alpha \tag{3.A.1}$$

and

$$\Pr\left[F_3(\alpha) \le \frac{\hat{\bar{\sigma}}_2^2\sigma_6^2}{\hat{\sigma}_\nu^2\sigma_5^2} \le F_4(\alpha)\right] = 1-\alpha \quad . \tag{3.A.2}$$

Recognizing that $(\sigma_4^2/\sigma_6^2) = 1 + T[\sigma_\mu^2(1-\rho_4) + \sigma_\nu^2\rho_7(1-\rho_6)]/\sigma_6^2 = 1 + T\bar{\rho}_3/(1-\bar{\rho}_3)$ and $(\sigma_5^2/\sigma_6^2) = 1 + n[\sigma_\tau^2(1-\rho_5) + \sigma_\nu^2\rho_6(1-\rho_7)]/\sigma_6^2 = 1 + n\bar{\rho}_2/(1-\bar{\rho}_2)$ we can write (3.A.1) and (3.A.2) as

$$\Pr\left[\frac{\hat{\bar{\sigma}}_1^2}{T\hat{\sigma}_\nu^2F_2(\alpha)} - \frac{1}{T} \le \frac{\bar{\rho}_3}{(1-\bar{\rho}_3)} \le \frac{\hat{\bar{\sigma}}_1^2}{T\hat{\sigma}_\nu^2F_1(\alpha)} - \frac{1}{T}\right] = 1-\alpha \tag{3.A.3}$$

and

$$\Pr\left[\frac{\hat{\bar{\sigma}}_2^2}{n\hat{\sigma}_\nu^2F_4(\alpha)} - \frac{1}{n} \le \frac{\bar{\rho}_2}{(1-\bar{\rho}_2)} \le \frac{\hat{\bar{\sigma}}_2^2}{n\hat{\sigma}_\nu^2F_3(\alpha)} - \frac{1}{n}\right] = 1-\alpha \tag{3.A.4}$$

respectively, where $\bar{\rho}_2$ and $\bar{\rho}_3$ are as explained in (3.12.13). The interval statements (3.A.3) and (3.A.4) imply that

$$\Pr\left[\frac{\hat{\bar{\sigma}}_1^{\,2} - F_2(\alpha)\hat{\sigma}_\upsilon^{\,2}}{\hat{\bar{\sigma}}_1^{\,2} + (T-1)F_2(\alpha)\hat{\sigma}_\upsilon^{\,2}} \le \bar{\rho}_3 \le \frac{\hat{\bar{\sigma}}_1^{\,2} - F_1(\alpha)\hat{\sigma}_\upsilon^{\,2}}{\hat{\bar{\sigma}}_1^{\,2} + (T-1)F_1(\alpha)\hat{\sigma}_\upsilon^{\,2}}\right] = 1-\alpha \qquad (3.A.5)$$

and

$$\Pr\left[\frac{\hat{\bar{\sigma}}_2^{\,2} - F_4(\alpha)\hat{\sigma}_\upsilon^{\,2}}{\hat{\bar{\sigma}}_2^{\,2} + (n-1)F_4(\alpha)\hat{\sigma}_\upsilon^{\,2}} \le \bar{\rho}_2 \le \frac{\hat{\bar{\sigma}}_2^{\,2} - F_3(\alpha)\hat{\sigma}_\upsilon^{\,2}}{\hat{\bar{\sigma}}_2^{\,2} + (n-1)F_3(\alpha)\hat{\sigma}_\upsilon^{\,2}}\right] = 1-\alpha \quad . \qquad (3.A.6)$$

CHAPTER IV

STATISTICAL INFERENCE IN RANDOM COEFFICIENT REGRESSION MODELS USING PANEL DATA

4.1 Introduction

In this chapter we modify Rao's statistical theory to suit our study of non-experimental situations. We restrict ourselves to situations where we have panel data. We specify a regression equation with coefficients random across units but coming from the same multivariate distribution. We develop an estimation procedure which produces at least asymptotically efficient estimators of the parameters of the model.

In Section 4.2 we pose the problem formally. Section 4.3 is devoted to the development of appropriate statistical inference techniques. In Section 4.4 we study the inference procedures which are appropriate when the disturbances are autocorrelated. Problems associated with the estimation of RCR models using aggregated data are considered in Section 4.5. We discuss optimal forecasting properties of RCR models and a comparison with those of fixed coefficient models in Section 4.6. A procedure of relaxing one of the assumptions underlying RCR models is indicated in Section 4.7. RCR methods are compared with Bayesian methods in Section 4.8. We indicate in Section 4.9 that the procedures developed in Section 4.3 are useful in estimating a production function with managerial input. In Section 4.10 we analyze a regression model with both fixed and random coefficients. We furnish conclusions of the study in Section 4.11.

4.2 Setting the Problem

(a) A RCR Model

Consider the following model.

$$\underline{y}_i = X_i \underline{\beta}_i + \underline{u}_i \qquad (i=1,2,\ldots,n) \ . \tag{4.2.1}$$

The vectors $\underline{y}_i$, X_i and $\underline{u}_i$ are as explained in (2.4.1). There are T observations on each of n individual units. Observed are $\underline{y}_i$ and X_i for i=1,2,...,n. X_i is a matrix of observations on K independent variables, x_{tki} (t=1,2,...,T; k=1,2,...,K). $\underline{\beta}_i$ and $\underline{u}_i$ are unobserved random vectors.

Unless otherwise specified the following assumption will be accepted as valid throughout this chapter.

Assumption 4.2.1:

(1) Same as (1) in Assumption 2.6.1.

(2) Same as (2) in Assumption 2.4.1.

(3) The disturbance vectors $\underline{u}_i$ (i=1,2,...,n) are independently distributed with the same mean vector zero. The variance-covariance matrix of $\underline{u}_i$ is $\sigma_{ii}I_T$.

(4) The coefficient vectors $\underline{\beta}_i$ (i=1,2,...,n) are independently and identically distributed with $E\underline{\beta}_i = \bar{\underline{\beta}}$ and $E(\underline{\beta}_i-\bar{\underline{\beta}})(\underline{\beta}_i-\bar{\underline{\beta}})'=\Delta$ which is nonsingular.

(5) The vectors $\underline{u}_i$ and $\underline{\beta}_j$ are independent for every i & j=1,2,...,n.

The model (4.2.1) represents a temporal cross-section situation with the subscript t representing time and the subscript i representing a micro unit. The specification (3) in Assumption 4.2.1 implies that the disturbances are both contemporaneously and serially uncorrelated, but the disturbances with different i subscript have different variances. The specification (4) in Assumption 4.2.1 implies that the regression coefficient vectors $\underline{\beta}_i$ (i=1,2,...,n) are random drawings from the same nonsingular multivariate distribution with mean $\bar{\underline{\beta}}$ and variance-covariance matrix Δ.

The parameters to be estimated are $\bar{\underline{\beta}}$, Δ and σ_{ii} (i=1,2,...,n). Now the problem is to obtain estimators with optimal properties for these parameters and to develop criteria of testing certain linear hypotheses on these parameters.

(b) An Example

Following Mundlak (1961) we can generalize CES production function. Let

$$V_{it} = \gamma[\delta K_{it}^{\rho_i} + (1-\delta)L_{it}^{\rho_i}]^{1/\rho_i} M_i^{\alpha} \tag{4.2.1a}$$

where V_{it} is an observation on the value added by the ith firm in the year t, K_{it} is an observation on the capital services, L_{it} is an observation on the labor services and M_i is an unobserved managerial input. We assume that the managerial input is invariant over time. γ, δ, ρ and α are fixed parameters. We write

$$V_{it}^{\rho_i} = \delta(\gamma M_i^{\alpha})^{\rho_i} K_{it}^{\rho_i} + (1-\delta)(\gamma M_i^{\alpha})^{\rho_i} L_{it}^{\rho_i}$$

or

$$\frac{V_{it}^{\rho_i}-1}{\rho_i} = \frac{(\gamma M_i^{\alpha})^{\rho_i}-1}{\rho_i} + \delta(\gamma M_i^{\alpha})^{\rho_i}\left(\frac{K_{it}^{\rho_i}-1}{\rho_i}\right) + (1-\delta)(\gamma M_i^{\alpha})^{\rho_i}\left(\frac{L_{it}^{\rho_i}-1}{\rho_i}\right) .$$

In terms of our notation in (4.2.1) we can write the above equation as

$$y_{it} = \beta_{0i} + \beta_{1i}x_{1it} + \beta_{2i}x_{2it} + u_{it} \tag{4.2.1b}$$

$$(i=1,2,\ldots,n;\ t=1,2,\ldots,T)$$

where $y_{it} = (V_{it}^{\rho_i}-1)/\rho_i$, $\beta_{0i} = [(\gamma M_i^{\alpha})^{\rho_i}-1]/\rho_i$, $\beta_{1i} = \delta(\gamma M_i^{\alpha})^{\rho_i}$, $x_{1it} = (K_{it}^{\rho_i}-1)/\rho_i$, $\beta_{2i} = (1-\delta)(\gamma M_i^{\alpha})^{\rho_i}$ and $x_{2it} = (L_{it}^{\rho_i}-1)/\rho_i$. We assume that the β's vary randomly across units. We indicate a procedure of estimating (4.2.1b) at the end of this chapter.

4.3 Efficient Methods of Estimating the Parameters of RCR Models

(a) Aitken Estimator of $\bar{\underline{\beta}}$

We can write, à la Zellner (1966),

$$\underline{\beta}_i = \bar{\underline{\beta}} + \underline{\delta}_i \qquad (i=1,2,\ldots,n) \tag{4.3.1}$$

where $\underline{\delta}_i$ is a KX1 vector of random elements. Now the specification (4) in Assumption 4.2.1 implies that $\underline{\delta}_i$ (i=1,2,...,n) are identically and independently distributed with mean zero and variance-covariance matrix Δ. We rewrite the equation (4.2.1) as

$$\begin{bmatrix} \underline{y}_1 \\ \underline{y}_2 \\ \cdot \\ \cdot \\ \cdot \\ \underline{y}_n \end{bmatrix} = \begin{bmatrix} X_1 \\ X_2 \\ \cdot \\ \cdot \\ \cdot \\ X_n \end{bmatrix} \underline{\bar{\beta}} + \begin{bmatrix} X_1 & 0 & \cdots & 0 \\ 0 & X_2 & \cdots & 0 \\ \cdot & \cdot & \cdots & \cdot \\ \cdot & \cdot & \cdots & \cdot \\ \cdot & \cdot & \cdots & \cdot \\ 0 & 0 & \cdots & X_n \end{bmatrix} \begin{bmatrix} \underline{\delta}_1 \\ \underline{\delta}_2 \\ \cdot \\ \cdot \\ \cdot \\ \underline{\delta}_n \end{bmatrix} + \begin{bmatrix} \underline{u}_1 \\ \underline{u}_2 \\ \cdot \\ \cdot \\ \cdot \\ \underline{u}_n \end{bmatrix} \tag{4.3.2}$$

or more compactly as,

$$\underline{y} = X\underline{\bar{\beta}} + D(X)\underline{\delta} + \underline{u} \tag{4.3.2a}$$

where $\underline{y} \equiv [\underline{y}_1', \underline{y}_2', \ldots, \underline{y}_n']'$, $X \equiv [X_1', X_2', \ldots, X_n']'$, $\underline{\delta} \equiv [\underline{\delta}_1', \underline{\delta}_2', \ldots, \underline{\delta}_n']'$, $\underline{u} \equiv [\underline{u}_1', \underline{u}_2', \ldots, \underline{u}_n']'$, the 0's are TXK null matrices and D(X) denotes the block-diagonal matrix on the r.h.s. of (4.3.2). Under Assumption 4.2.1 the nTX1 disturbance vector $D(X)\underline{\delta}+\underline{u}$ has the following variance-covariance matrix.

$$H(\underline{\theta}) = \begin{bmatrix} X_1\Delta X_1' + \sigma_{11}I_T & 0 & \cdots & 0 \\ 0 & X_2\Delta X_2' + \sigma_{22}I_T & \cdots & 0 \\ \cdot & \cdot & & \cdot \\ \cdot & \cdot & & \cdot \\ \cdot & \cdot & & \cdot \\ 0 & 0 & \cdots & X_n\Delta X_n' + \sigma_{nn}I_T \end{bmatrix} \tag{4.3.3}$$

where the 0's are TxT null matrices. $H(\underline{\theta})$ is a symmetric nTxnT matrix which is a function of the X_i's and an unknown $\frac{1}{2}[K(K+1) + 2n]$- element parametric vector $\underline{\theta}$ containing all the distinct elements of Δ and σ_{ii} (i=1,2,...,n) arranged in an order. We assume that each element of $H(\underline{\theta})$ has continuous first order derivatives. In some $\frac{1}{2}[K(K+1) + 2n]$- dimensional interval A_1 that contains the true value of $\underline{\theta}$, say $\underline{\theta}_0$, as an interior point, $H(\underline{\theta})$ is positive definite for all $\underline{\theta}\epsilon A_1$.

To estimate $\bar{\underline{\beta}}$ we apply Aitken's generalized least squares to (4.3.2a). Thus the BLUE of $\bar{\underline{\beta}}$ is

$$\bar{\underline{b}}(\underline{\theta}) = (X'H(\underline{\theta})^{-1}X)^{-1}X'H(\underline{\theta})^{-1}\underline{y}$$

$$= \left[\sum_{j=1}^{n} X_j' \left\{X_j \Delta X_j' + \sigma_{jj} I_T\right\}^{-1} X_j\right]^{-1}$$

$$\cdot\left[\sum_{i=1}^{n} X_i' \left\{X_i \Delta X_i' + \sigma_{ii} I_T\right\}^{-1} \underline{y}_i\right] \quad . \qquad (4.3.4)$$

We now make use of a matrix result given in Rao (1965a, p. 29, Problem 2.9) to show that

$$(\sigma_{ii} I_T + X_i \Delta X_i')^{-1} = \frac{M_i}{\sigma_{ii}} + X_i (X_i'X_i)^{-1} \left\{\Delta + \sigma_{ii}(X_i'X_i)^{-1}\right\}^{-1} (X_i'X_i)^{-1} X_i' \quad (4.3.5)$$

where $M_i = I_T - X_i(X_i'X_i)^{-1}X_i'$. This result can be verified by premultiplying both sides by $(\sigma_{ii} I_T + X_i \Delta X_i')$.

Substituting (4.3.5) in (4.3.4) we have

$$\bar{\underline{b}}(\underline{\theta}) = \sum_{i=1}^{n} W_i(\underline{\theta}) \underline{b}_i \qquad (4.3.6)$$

where $W_i(\underline{\theta}) = \left[\sum_{j=1}^{n} \left\{\Delta + \sigma_{jj}(X_j'X_j)^{-1}\right\}^{-1}\right]^{-1} \left\{\Delta + \sigma_{ii}(X_i'X_i)^{-1}\right\}^{-1}$, and $\underline{b}_i = (X_i'X_i)^{-1} \cdot X_i'\underline{y}_i$.

The variance-covariance matrix of the estimator $\bar{\underline{b}}(\underline{\theta})$ is easily shown to be $(X'H(\underline{\theta})^{-1}X)^{-1}$ or

$$V[\bar{\underline{b}}(\underline{\theta})] = C(\underline{\theta}) = \left[\sum_{i=1}^{n} X_i' \left\{X_i \Delta X_i' + \sigma_{ii} I\right\}^{-1} X_i\right]^{-1} \qquad (4.3.7)$$

$$= \left[\sum_{i=1}^{n} \left\{\Delta + \sigma_{ii}(X_i'X_i)^{-1}\right\}^{-1}\right]^{-1} \quad . \qquad (4.3.8)$$

It is to be noted that (4.3.6) is equal to the estimator (1.2.38) suggested by Rao (1965b) if $X_1 = X_2 = \ldots = X_n$ and $\sigma_{11} = \sigma_{22} = \ldots = \sigma_{nn}$. More generally, the estimator (4.3.6) will be recognized as the weighted average of the estimators

$\underline{b}_i$(i=1,2,...,n) which can be obtained by applying least squares method to each equation separately in (4.2.1). According to Rao's (1965b) lemma 1.2.1 the BLU predictor of $\underline{\beta}_i$ in the ith equation of the type (4.2.1) reduces to

$$\underline{b}_i = (X_i'X_i)^{-1}X_i'\underline{y}_i \tag{4.3.9}$$

which is a linear function of $\underline{y}_i$.

The error of $\underline{b}_i$ in predicting $\underline{\beta}_i$ is given by

$$\underline{b}_i - \underline{\beta}_i = (X_i'X_i)^{-1} X_i'\underline{u}_i \quad . \tag{4.3.10}$$

This prediction error has expectation, $E(\underline{b}_i-\underline{\beta}_i)=0$ and variance-covariance matrix

$$E(\underline{b}_i-\underline{\beta}_i)(\underline{b}_i-\underline{\beta}_i)' = \sigma_{ii}(X_i'X_i)^{-1} \quad . \tag{4.3.11}$$

According to Rao's (1965b) lemma 1.2.2 $\underline{b}_i$ is also a best linear -- in $\underline{y}_i$ -- unbiased estimator of $\underline{\bar{\beta}}$.

Alternatively, we can show that the BLUE of $\underline{\bar{\beta}}$ in the ith equation, $\underline{y}_i = X_i\underline{\bar{\beta}} + X_i\underline{\delta}_i + \underline{u}_i$ is given by the Aitken estimator

$$\underline{b}_i = [X_i'(X_i\Delta X_i' + \sigma_{ii}I)^{-1}X_i]^{-1}X_i'(X_i\Delta X_i' + \sigma_{ii}I)^{-1}\underline{y}_i \quad . \tag{4.3.12}$$

Using (4.3.5) we obtain

$$\underline{b}_i = [\Delta + \sigma_{ii}(X_i'X_i)^{-1}]\,[\Delta + \sigma_{ii}(X_i'X_i)^{-1}]^{-1}(X_i'X_i)^{-1}X_i'\underline{y}_i = (X_i'X_i)^{-1}X_i'\underline{y}_i \ .$$

Thus in this particular situation the Aitken estimator (4.3.12) is the same as the OLS estimator, $\underline{b}_i$. This is so because the condition (2.3.4b) is satisfied in this case. If we have a micro unit whose behavior is explained by an equation of the type (4.2.1), then the OLS estimator is the BLUE of the mean of the coefficient vector but the conventional formula for calculating the standard errors is inappropriate. This can be shown as below. The error of $\underline{b}_i$ as an estimator of $\underline{\bar{\beta}}$ is

$$\underline{b}_i - \underline{\bar{\beta}} = (\underline{b}_i-\underline{\beta}_i) + (\underline{\beta}_i-\underline{\bar{\beta}}) \quad .$$

This sampling error has expectation,

$$E(\underline{b}_i - \overline{\underline{\beta}}) = 0 \qquad (4.3.13)$$

and variance-covariance matrix,

$$E(\underline{b}_i - \overline{\underline{\beta}})(\underline{b}_i - \overline{\underline{\beta}})' = \Delta + \sigma_{ii}(X_i'X_i)^{-1} \quad . \qquad (4.3.14)$$

For our panel of nT observations we have a set of n functions $\underline{b}_i$ $(i=1,2,\ldots,n)$. Neither the variance of an element of the prediction error (4.3.10) nor the variance of an element of the estimator $\underline{b}_i$ is the same for all $\underline{b}_i$'s because the X_i's differ among individual units.

Notice here that the $\underline{b}_i$ is at the same time a best predictor of $\underline{\beta}_i$ and a best estimator of $\overline{\underline{\beta}}$. But its variance-covariance matrix as an estimator of $\overline{\underline{\beta}}$ is not the same as its variance-covariance matrix as a predictor of $\underline{\beta}_i$. The variance of an element of $\underline{b}_i$ as an estimator is larger than its variance as a predictor. In this connection it is essential to note that we are considering a linear function of $\underline{y}_i$ as "predicting" the drawing on $\underline{\beta}_i$ which produced that $\underline{y}_i$, not as predicting a fresh drawing on $\underline{\beta}_i$ nor a $\underline{\beta}_j$ for $i \neq j$. If we are going to use $\underline{y}_i$ to really predict either a fresh drawing on $\underline{\beta}_i$ or for that matter a $\underline{\beta}_j$ $(i\neq j)$, $\underline{b}_i$ would yield a prediction error with zero expectation and variance-covariance matrix $2\Delta + \sigma_{ii} \cdot (X_i'X_i)^{-1}$, whose diagonal elements are larger than the corresponding diagonal elements of the variance-covariance matrix in (4.3.11).

Since the $\underline{\beta}_i$'s are identically distributed vectors, the n functions $\underline{b}_i$ provide n different linear unbiased and uncorrelated estimators (with unequal variances) of the same parametric vector $\overline{\underline{\beta}}$. The best way to pool them into a single estimator is to take a weighted average of all n estimators with weights inversely proportional to their variances. This is what is done in (4.3.6). That $\overline{\underline{b}}(\theta)$ has minimum variance among all unbiased linear combinations of $\underline{b}_i$'s is attributable to the fact that it minimizes the quadratic form

$$\sum_{i=1}^{n} (\underline{b}_i - \overline{\underline{\beta}})' [\Delta + \sigma_{ii}(X_i'X_i)^{-1}]^{-1} (\underline{b}_i - \overline{\underline{\beta}}) \quad . \qquad (4.3.15)$$

The criterion of minimizing (4.3.15) is identical to that of the generalized least squares principle applied to the $\underline{b}_i$'s as observations. The concentration ellipsoid of $\underline{\bar{b}}(\underline{\theta})$ is contained in that of every other linear unbiased estimator of $\underline{\bar{\beta}}$, cf. Malinvaud (1966, p. 149).

Here the distinction between $\underline{b}_i$ as the BLUE of $\underline{\bar{\beta}}$ and $\underline{\bar{b}}(\underline{\theta})$ as the BLUE of the same $\underline{\bar{\beta}}$ is to be clearly understood. What we are seeking through Rao's lemma 1.2.2 is a linear function of $\underline{y}_i$, namely $\underline{b}_i$ which is the BLUE of $\underline{\bar{\beta}}$ in the sense that any other estimator of $\underline{\bar{\beta}}$, which is also linear in the vector $\underline{y}_i$ and unbiased has a variance-covariance matrix which exceeds that of $\underline{b}_i$ by a positive semidefinite matrix. On the other hand, $\underline{\bar{b}}(\underline{\theta})$ is the BLUE of $\underline{\bar{\beta}}$ in the sense that any other estimator of $\underline{\bar{\beta}}$, which is also linear in the vector $\underline{y}$, given in (4.3.2a), and unbiased has a variance-covariance matrix which exceeds that of $\underline{\bar{b}}(\underline{\theta})$ by a positive semidefinite matrix.

If Δ and σ_{ii} are unknown, as they usually are, it is not possible to use (4.3.6) in practice. A feasible procedure is to employ estimators of Δ and σ_{ii} in constructing an estimator of the type (4.3.6). In view of our discussion in subsection 2.4(c) and in Sections 3.6 and 3.7, such a procedure is desirable only if it leads to an estimator which is better than an OLS estimator or any other estimator in a minimum variance sense. Since X is a nTxK matrix of rank K, we can find a nTx(nT-K) matrix, say L, of rank nT-K such that $L'X=0$. We can write $L \equiv [L_1', L_2', \ldots, L_n']'$ such that each L_i is a Tx(nT-K) matrix and $L'X = \sum_{i=1}^{n} L_i'X_i = 0$. Premultiplying (4.3.3) by X' and postmultiplying by L we have

$$X'H(\underline{\theta})L = \sum_{i=1}^{n} (X_i'X_i\Delta X_i'L_i + \sigma_{ii}X_i'L_i) \quad . \qquad (4.3.16)$$

Unless Δ is close to a null matrix and σ_{ii} is close to σ^2 for every i, $X'H(\underline{\theta})L$ cannot be close to a null matrix. Our discussion in Section 3.7 indicates that if $X'H(\underline{\theta})L$ is appreciably different from null, an Aitken estimator of $\underline{\bar{\beta}}$ based on an estimator of $\underline{\theta}$ will be more efficient than an OLS estimator of $\underline{\bar{\beta}}$ even in small samples. We will show in the subsequent sections that an Aitken estimator of $\underline{\bar{\beta}}$ based on the consistent estimates of Δ and σ_{ii}'s will be asymptotically as efficient as the Aitken estimator $\bar{\underline{b}}(\underline{\theta})$ in (4.3.4). We also develop an asymptotic procedure to test the hypothesis $\Delta=0$. This procedure will be useful in testing whether $X'H(\underline{\theta})L$ is close to null. If $X'H(\underline{\theta})L$ is close to null, OLS estimator of $\bar{\underline{\beta}}$ will be more efficient than an Aitken estimator based on an estimate of $\underline{\theta}$ in small samples.

(b) Comparison of Random Coefficient Models with Fixed Coefficient Models

We digress here from the main theme of the chapter and consider in this subsection a fixed coefficient model.

Suppose we treat the model in (4.2.1) as a fixed coefficient model and estimate it by pooling all the nT observations on variables and imposing the conditions, $\underline{\beta}_1 = \underline{\beta}_2 = \ldots = \underline{\beta}_n = \underline{\beta}$, say, and $\sigma_{11} = \sigma_{22} = \ldots = \sigma_{nn} = \sigma^2$, say. The BLUE of $\underline{\beta}$ under these assumptions is

$$\underline{b} = \left(\sum_{i=1}^{n} X_i'X_i \right)^{-1} \left(\sum_{i=1}^{n} X_i'\underline{y}_i \right) \quad . \tag{4.3.17}$$

$\underline{b}$ can be viewed as a weighted average of $\underline{b}_i$ (i=1,2,...,n) where the weight given to $\underline{b}_i$ is equal to $\left(\sum_{i=1}^{n} X_i'X_i \right)^{-1} X_i'X_i$. It is instructive to compare these weights with those in (4.3.6).

It can easily be shown that

$$E\underline{b} = \underline{\beta} \tag{4.3.18}$$

and the variance-covariance of $\underline{b}$ is

$$V(\underline{b}) = \sigma^2 \left(\sum_{i=1}^{n} X_i'X_i \right)^{-1} \quad . \tag{4.3.19}$$

An unbiased estimate of σ^2 is given by

$$\sum_{i=1}^{n} \frac{(\underline{y}_i - X_i\underline{b})'(\underline{y}_i - X_i\underline{b})}{nT-K} \quad .$$

If the regression coefficient vectors $\underline{\beta}_i$ are random across units but we misspecify them as a fixed coefficient vector, we will be estimating $\bar{\underline{\beta}}$ by $\underline{b}$, rather than by $\bar{\underline{b}}(\underline{\theta})$. The variance formula in (4.3.19) and the estimate of σ^2 are inappropriate, though $E\underline{b}$ is still equal to $\bar{\underline{\beta}}$.

The variance-covariance matrix of $\underline{b}$ when $\underline{\beta}_i$'s are random, is given by

$$V(\underline{b}) = \left(\sum_{i=1}^{n} X_i'X_i\right)^{-1}\left(\sum_{i=1}^{n} X_i'X_i\Delta X_i'X_i\right) \cdot \left(\sum_{i=1}^{n} X_i'X_i\right)^{-1} + \sigma^2\left(\sum_{i=1}^{n} X_i'X_i\right)^{-1} \quad . \tag{4.3.20}$$

Using (4.3.20) we can obtain the correct variance of the OLS estimator (4.3.17) if we know Δ and σ^2.

If we treat (4.2.1) as a fixed coefficient model and assume that each individual has a separate coefficient vector, then a single macro relation cannot be derived by aggregation from a set of micro relations in (4.2.1). Data on each individual unit should be used to estimate its own coefficient vector. The number of coefficient vectors to be estimated grows with the number of individual units in the sample. We can get around this difficulty if we can treat the coefficient vector in (4.2.1) as a random variable changing across individuals. When we go from a model with fixed parameters varying systematically across units to a RCR model, the number of unknown fixed parameters is reduced substantially which is usually desirable. For example, if we treat (4.2.1) as a fixed coefficient model with coefficients differing from one individual to another we will have Kn unknown regression coefficients to be estimated and this number increases with n. In the case of a RCR model of (4.3.2) type there are only K unknown elements of the mean vector and $\frac{1}{2}$ [K(K+1)] distinct elements of the variance-covariance matrix. The proportion of sample information used to estimate unknown parameters appearing in a RCR model decreases toward zero as the sample size n increases.

(c) Estimation of the Means of Random Coefficients when Δ and σ_{ii} are Unknown

Now we continue the study of RCR models. We first make an attempt to find unbiased estimators of Δ and σ_{ii}.

An unbiased estimator of σ_{ii} which is the variance of the ith units disturbance term for any time period, is given by

$$s_{ii} = \frac{\underline{u}_i'M_i\underline{u}_i}{T-K} \tag{4.3.21}$$

where M_i is as shown in (4.3.5). It can be easily verified that the estimator s_{ii} which is based on $M_i\underline{u}_i=\hat{\underline{u}}_i$, the vector of least squares residuals from a least squares fit of $\underline{y}_i$ upon X_i, has an expectation equal to σ_{ii}.

To estimate Δ we treat the least squares quantities $\underline{b}_i(i=1,2,\ldots,n)$, as a random sample of size n. $\underline{b}_i$, the ith drawing in this sample, is the sum of two random vectors, namely $\underline{\beta}_i$ and a linear function of $\underline{u}_i$. Now define

$$S_b = \sum_{i=1}^{n} \underline{b}_i\underline{b}_i' - \frac{1}{n}\sum_{i=1}^{n}\underline{b}_i\sum_{i=1}^{n}\underline{b}_i' \quad . \tag{4.3.22}$$

$\frac{S_b}{n-1}$ is the sample variance-covariance matrix of the $\underline{b}_i$.

Taking expectations on both sides of (4.3.22) after substituting $\underline{\beta}_i + (X_i'X_i)^{-1}X_i'\underline{u}_i$ for $\underline{b}_i$ we have

$$ES_b = n(\Delta + \underline{\bar{\beta}}\,\underline{\bar{\beta}}') + \sum_{i=1}^{n}\sigma_{ii}(X_i'X_i)^{-1} - (\Delta + \underline{\bar{\beta}}\,\underline{\bar{\beta}}') - (n-1)\underline{\bar{\beta}}\,\underline{\bar{\beta}}'$$

$$- \frac{1}{n}\sum_{i=1}^{n}\sigma_{ii}\,(X_i'X_i)^{-1}$$

$$= (n-1)\Delta + \frac{(n-1)}{n}\sum_{i=1}^{n}\sigma_{ii}(X_i'X_i)^{-1} \quad . \tag{4.3.23}$$

Therefore,

$$\hat{\Delta} = \frac{S_b}{n-1} - \frac{1}{n}\sum_{i=1}^{n} s_{ii}\,(X_i'X_i)^{-1} \tag{4.3.24}$$

is an unbiased estimator of Δ. This result critically depends upon Assumption 4.2.1.

Before moving on to the next subsection we point out some fundamental problems associated with the estimator $\hat{\Delta}$. Since Δ is a variance-covariance matrix and it is assumed to be nonsingular, all its diagonal elements should be positive and its i-jth off diagonal element, when squared, should at-most be equal to the product of its ith and jth diagonal elements. Since these restrictions are neglected in our estimation procedure, the elements of $\hat{\Delta}$ may violate these restrictions with positive probability. In some numerical applications $\hat{\Delta}$ will yield negative estimates for the variances of some coefficients. This difficulty is familiar in discussions of random-effect models in the analysis of variance. These models can be represented as $y_{it} = \mu + a_i + u_{it}$ $(i=1,2,\ldots,n;\ t=1,2,\ldots,T)$ where the a_i's and the u_{it}'s are independently distributed random variables such that $Ea_i = 0$, $Ea_i^2 = \sigma_a^2$, $Eu_{it} = 0$ and $Eu_{it}^2 = \sigma^2$ for all i and t. An unbiased estimator of σ_a^2 is $\hat{\sigma}_a^2 = B/T(n-1) - W/Tn(T-1)$ where B and W are the sums of squares between and within groups, respectively. However this estimate can assume negative values with positive probability. Although the model (4.2.1) differs from the random-effect models in many ways, there are similarities between the difficulties associated with $\hat{\Delta}$ and those with $\hat{\sigma}_a^2$. We may gain some insights into the difficulty associated with $\hat{\Delta}$ if we closely study the unbiased estimator of σ_a^2. Therefore we briefly review the literature on random-effect models and mention the solutions offered by different people to the negative estimated variance problem associated with $\hat{\sigma}_a^2$.

Scheffé (1959, p. 229) has suggested setting the variance equal to zero whenever a negative estimate is obtained in a random-effect model. On the other hand, if one attempts to restrict the value of the variance estimator to be nonnegative, this will destroy its unbiasedness property, cf. [Herbach (1959), Thompson (1962, 1963)].

According to Thompson two possible explanations of a negative estimate are: (1) the assumed model may be incorrect and (2) statistical noise may have obscured the underlying physical situation. McHugh and Mielke (1968) point out that statistical dependence of a_i and a_j for $i \neq j$, and of u_{it} and $u_{it'}$ for $t \neq t'$ is a consequence of non-replacement sampling from an existent population which is necessarily finite. If this is true and if the assumption of statistical independence of a_i and a_j $(i \neq j)$, and of u_{it} and $u_{it'}$ $(t \neq t')$ is in error, then the usual variance estimator does not necessarily possess a nonnegative expectation. That is, when the usual specification of statistically independent random variable is false, negativity of the expectation of the usual variance estimator is a well defined possibility with the result that negative estimates of variances cannot be excluded on theoretical grounds. We obtained a similar result in subsection 3.12(a). The above argument indicates a situation for which the specification (3) in Assumption 3.12.1 is valid. If the cross-section units are randomly drawn from a finite population of units without replacement, we can show that the covariance between μ_i and μ_j (with $i \neq j$) is $-\sigma_\mu^2/N_p$ where N_p is the number of units in the population. In this case we can adopt the specification (3) in Assumption 3.12.1 and restrict ρ_4 to be equal to $-1/N_p$.

Tiao and Tan (1965) have utilized Bayesian methods to analyze a random-effect model and have concluded that the negative estimated variance problem does not exist when one analyzes a random-effect model from a Bayesian point of view. This result is the consequence of putting in from the Bayesian point of view the prior information that variances cannot be negative. Tiao and Tan have shown that the situations in which the traditional unbiased estimator of σ_a^2 assumes a negative value will correspond in a Bayesian argument to a posterior distribution of σ_a^2 having its mode at the origin. That is, when the classical unbiased estimator of σ_a^2 assumes a negative value, the posterior distribution of σ_a^2 is J-shaped, having its maximum at the origin, and rapidly decreasing toward the right. Since these posterior distributions are obtained by taking non-informative prior distributions determined by Jeffrey's invariance criterion, this implies that a relatively more weight to small values of the variance in the posterior than in the prior is given and this is presumably in accord with the practice of some frequentists who set the variance equal to zero whenever its estimate is negative.

B. M. Hill (1965) also considered the estimation of variance components (σ_a^2 and σ^2) in the one-way random-effect model from a Bayesian point of view. He has argued that a large negative estimate for σ_a^2 indicates an uninformative experiment in which the effective likelihood for that variance component is extremely flat. In such circumstances the posterior distribution depends critically upon the prior and any conventional non-informative prior introduces arbitrariness in posterior inferences. According to Hill the posterior distributions derived by Tiao and Tan taking the invariance diffuse priors are inappropriate as measures of posterior opinion when the traditional unbiased estimator of the variance component assumes a large negative value. As is done by Hill proper prior distributions are to be introduced whenever the effective likelihood for a variance component is flat. When an informative experiment is performed in the sense that the likelihood function is quite sharp, it makes little difference what the exact shape of prior is because within a narrow range where the likelihood takes substantial values any prior can well be approximated by a rectangular distribution. In such cases certain conventional diffuse priors are convenient although largely arbitrary. Stone and Springer (1965) pointed out a paradox involving quasi prior distributions adopted by Tiao and Tan.

In another paper, Tiao and Tan (1966) utilized Bayesian methods to analyze the model $y_{it} = \mu + a_i + u_{it}$ in which the errors u_{it} were assumed to follow a first order Markov process $u_{it} = \rho u_{it-1} + \epsilon_{it}$. It was shown that inferences about the variances σ_a^2 and σ^2 were sensitive to changes in the value of ρ assumed. In particular, in a set of data generated from a population with $\rho = -0.5$, $\sigma = 1.5$ and $\sigma_a^2 = 1$, they found that when the independence assumption $\rho = 0$ was made, the classical unbiased estimator of σ_a^2 assumed a negative value. A large deviation of assumed value for ρ from the true value of ρ would tend to overestimate σ^2. The extent of bias in the estimates of σ^2 and σ_a^2 caused by the discrepancy between the true value of ρ and the assumed value of ρ was particularly alarming when the u_{it}'s were in fact highly negatively autocorrelated and was relatively mild in the case of positive auto-correlation. We have shown in subsection 3.12(a) that this result is true even when the error components are equicorrelated.

Thus an alternative explanation for a negative estimated variance, apart from chance, could be misspecification of the basic regression model. Scheffé's approach of setting the variance equal to zero whenever a negative estimate is obtained or the Tiao and Tan (1965) approach of constraining the range of variances to be positive may not be valid if the model is misspecified. Misspecification is more likely to happen in the analysis of non-experimental data than in that of experimental data. In the former case the negative variance estimates may possibly be associated with a misspecification of the model. This problem will be explored more fully in Section 4.10 below and in the subsequent chapters with the help of some numerical examples.

Now that we have unbiased estimators of Δ and σ_{ii} we can adopt the following estimator of $\underline{\bar{\beta}}$.

$$\underline{\bar{b}}(\underline{\hat{\theta}}) = \sum_{i=1}^{n} W_i \; (\underline{\hat{\theta}})\underline{b}_i \tag{4.3.25}$$

where $W_i(\underline{\hat{\theta}}) = C(\underline{\hat{\theta}}) \left\{\hat{\Delta} + s_{ii}(X_i'X_i)^{-1}\right\}^{-1}$, and $C(\underline{\hat{\theta}}) = \left[\sum_{j=1}^{n} \left\{\hat{\Delta} + s_{jj}(X_j'X_j)^{-1}\right\}^{-1}\right]^{-1}$.

We will show in the subsequent pages of this chapter that $\underline{\bar{b}}(\underline{\hat{\theta}})$ is a consistent, asymptotically normal and asymptotically efficient estimator with asymptotic variance-covariance matrix $\frac{\Delta}{n}$ as $T \to \infty$ and n is fixed. The estimated asymptotic variance-covariance matrix of $\underline{\bar{b}}(\underline{\hat{\theta}})$ is $\frac{\hat{\Delta}}{n}$.

(d) Maximum Likelihood Estimators of Parameters in the RCR Model

We shall now assume specific distributions for the variables involved.

Assumption 4.3.1: The disturbance vectors $\underline{u}_i$ (i=1,2,...,n) and the coefficient vectors $\underline{\beta}_i$ (i=1,2,...,n) are normally distributed.

Assumptions 4.2.1 and 4.3.1 together imply that the vector $\underline{y}_i$ is normally distributed with mean vector $X_i\underline{\bar{\beta}}$ and variance-covariance matrix $X_i\Delta X_i' + \sigma_{ii}I_T$, i.e., $\underline{y}_i \sim N_T(X_i\underline{\bar{\beta}},\; X_i\Delta X_i' + \sigma_{ii}I_T)$. The vectors $\underline{y}_1,\ldots,\underline{y}_n$ are mutually independent. Under these conditions $\underline{\bar{b}}(\underline{\theta})$ is an ML estimator, given the $\underline{y}_i$'s as observations because $\underline{\bar{b}}(\underline{\theta})$ minimizes the quadratic form in (4.3.15), which is a term in the exponent of the following likelihood function containing $\underline{\bar{\beta}}$ and $\underline{\theta}$.

$$L(\underline{\bar{\beta}},\underline{\theta}|\underline{y},X) = (2\pi)^{-\frac{nT}{2}} \prod_{i=1}^{n} |X_i\Delta X_i' + \sigma_{ii}I_T|^{-\frac{1}{2}}$$
$$\cdot \exp\left\{-\frac{1}{2}\sum_{i=1}^{n}(\underline{y}_i - X_i\underline{\bar{\beta}})'[X_i\Delta X_i' + \sigma_{ii}I_T]^{-1}(\underline{y}_i - X_i\underline{\bar{\beta}})\right\} . \tag{4.3.26}$$

We can write

$$|X_i\Delta X_i' + \sigma_{ii}I_T| = \begin{vmatrix} \sigma_{ii}I_T & X_i\Delta \\ -X_i' & I_K \end{vmatrix} = \sigma_{ii}^{T-K}|X_i'X_i||\Delta + \sigma_{ii}(X_i'X_i)^{-1}| . \tag{4.3.27}$$

Making use of results in (4.3.5), (4.3.9), (4.3.21) and (4.3.27) we can write the likelihood function as

$$L(\underline{\bar{\beta}},\underline{\theta}|\underline{y},X) = (2\pi)^{-\frac{nT}{2}} \prod_{i=1}^{n} \left\{\sigma_{ii}^{-\frac{1}{2}(T-K)} |X_i'X_i|^{-\frac{1}{2}} |\Delta + \sigma_{ii}(X_i'X_i)^{-1}|^{-\frac{1}{2}}\right\}$$

$$\cdot \exp\left\{-\frac{1}{2}\sum_{i=1}^{n}\left[\frac{(T-K)s_{ii}}{\sigma_{ii}} + (\underline{b}_i-\underline{\bar{\beta}})'[\Delta+\sigma_{ii}(X_i'X_i)^{-1}]^{-1}(\underline{b}_i-\underline{\bar{\beta}}\right]\right\}. \tag{4.3.28}$$

The differentiation of the log-likelihood with respect to $\underline{\bar{\beta}}$, σ_{ii}, and Δ yields the following equations:

$$\frac{\partial \log L}{\partial \underline{\bar{\beta}}} = \sum_{i=1}^{n} [\Delta + \sigma_{ii}(X_i'X_i)^{-1}]^{-1} (\underline{b}_i - \underline{\bar{\beta}}) = 0 \tag{4.3.29}$$

$$\frac{\partial \log L}{\partial \sigma_{ii}} = -\frac{1}{2}(T-K)/\sigma_{ii} - \frac{1}{2}\,\mathrm{tr}[\Delta + \sigma_{ii}(X_i'X_i)^{-1}]^{-1} (X_i'X_i)^{-1}$$

$$+\frac{1}{2}\frac{(T-K)s_{ii}}{\sigma_{ii}^2} + \frac{1}{2}\,\mathrm{tr}[\Delta + \sigma_{ii}(X_i'X_i)^{-1}]^{-1}$$

$$\cdot (\underline{b}_i-\underline{\bar{\beta}})(\underline{b}_i-\underline{\bar{\beta}})'[\Delta + \sigma_{ii}(X_i'X_i)^{-1}]^{-1} (X_i'X_i)^{-1} = 0 \tag{4.3.30}$$

$$(i=1,2,\ldots,n)$$

cf. [Dwyer (1967, 11.8), and Hartley and Rao (1967, 12)].

$$\frac{\partial \log L}{\partial \Delta} = -\frac{1}{2}\sum_{i=1}^{n} [\Delta + \sigma_{ii}(X_i'X_i)^{-1}]^{-1} + \frac{1}{2}\sum_{j=1}^{n} [\Delta + \sigma_{ii}(X_i'X_i)^{-1}]^{-1}$$

$$\cdot (\underline{b}_i-\underline{\bar{\beta}})(\underline{b}-\underline{\bar{\beta}})' [\Delta + \sigma_{ii}(X_i'X_i)^{-1}]^{-1} = 0, \tag{4.3.31}$$

cf. Dwyer (1967, 11.8).

By solving the equation (4.3.29) for $\underline{\bar{\beta}}$ in terms of $\underline{\theta}$ we obtain $\underline{\bar{b}}(\underline{\theta})$ in (4.3.6) as an ML estimator of $\underline{\bar{\beta}}$. The algebraic solution of equations (4.3.30) and (4.3.31) is quite difficult since the equations are nonlinear in the unknowns. Substitution of (4.3.6) in (4.3.30) and (4.3.31) would yield $n + K(K+1)/2$ equations in terms of the elements of $\underline{\theta}$, $\underline{y}$ and X. From these we obtain the equations

$$\frac{\partial \log L}{\partial \Delta} \frac{\partial \Delta}{\partial \sigma^2_{\beta_k}} \frac{\partial \sigma^2_{\beta_k}}{\partial \sigma_{\beta_k}} = 0 \qquad (4.3.32)$$

and

$$\frac{\partial \log L}{\partial \Delta} \frac{\partial \Delta}{\partial \sigma_{\beta_k \beta_{k'}}} = 0 \qquad \text{for } k \neq k' \qquad (4.3.33)$$

where $\sigma^2_{\beta_k}$ is the variance of k-th coefficient and $\sigma_{\beta_k \beta_{k'}} (k \neq k')$ is the covariance between a pair of coefficients. In other words $\sigma^2_{\beta_k}$ is the k-th diagonal element of Δ and $\sigma_{\beta_k \beta_{k'}}$ is an offdiagonal element of Δ. Using the steepest ascent method developed by Hartley and Rao (1967), we can solve the equations (4.3.30), (4.3.32) and (4.3.33), and obtain an ML estimate of $\underline{\theta}$. Substitution of ML estimate of $\underline{\theta}$ in (4.3.6) yields an ML estimate of $\underline{\bar{\beta}}$. Let $\underline{\tilde{\theta}}$ and $b(\underline{\tilde{\theta}})$ represent the ML estimates of $\underline{\theta}$ and $\underline{\bar{\beta}}$ respectively. We can form the ML estimates of σ_{ii} and Δ from the elements of $\underline{\tilde{\theta}}$. Let $\tilde{s}_{ii}$ and $\tilde{\Delta}$ be the ML estimates of σ_{ii} and Δ respectively.

The steepest ascent method is primarily developed to provide nonnegative estimates for variances. If we want to incorporate the restrictions on the covariances $\sigma_{\beta_k \beta_{k'}}$, we can adopt Nelder's (1968) transformation of the elements of Δ and constrain the ML estimate of Δ to be positive-(semi) definite.

Following Hartley and Rao (1967) we can show that under certain general conditions, if the starting values ${}_0\underline{\bar{\beta}}$ and ${}_0\underline{\theta}$ of the steepest ascent method are consistent estimators of the true parameters $\underline{\bar{\beta}}_0$ and $\underline{\theta}_0$ respectively, then the ML estimators $\underline{\bar{b}}(\underline{\tilde{\theta}})$ and $\underline{\tilde{\theta}}$ are consistent. If $\underline{\bar{b}}(\underline{\tilde{\theta}})$ and $\underline{\tilde{\theta}}$ provide the global maximum of the likelihood, then they are consistent.

(e) Minimum Variance Unbiased Estimator of $\underline{\bar{\beta}}$

In subsection 4.3(a) we have shown that $\underline{\bar{b}}(\underline{\theta})$ is the BLUE of $\underline{\bar{\beta}}$. Under the additional Assumption 4.3.1 we will show that $\underline{\bar{b}}(\underline{\theta})$ is a best estimator not only within the class of linear unbiased estimators, but also within the class of "regular" unbiased estimators.[17]

---

[17] The word "regular" here means that all the members of the class are functions of $\underline{y}$ whose density function satisfies the regularity conditions given in Cramér (1946, p. 479).

Differentiating (4.3.29) once again with respect to $\underline{\bar{\beta}}$, and taking expectation we obtain

$$- \sum_{i=1}^{n} [\Delta + \sigma_{ii}(X_i'X_i)^{-1}]^{-1} \quad . \qquad (4.3.34)$$

In the same way if we differentiate (4.3.30) and (4.3.31) with respect to $\underline{\bar{\beta}}$ we obtain

$$\frac{\partial^2 \log L}{\partial\sigma_{ii}\partial\underline{\bar{\beta}}} = - [\Delta + \sigma_{ii}(X_i'X_i)^{-1}]^{-1}(X_i'X_i)^{-1}[\Delta + \sigma_{ii}(X_i'X_i)^{-1}]^{-1}(\underline{b}_i - \underline{\bar{\beta}})$$

$$(i=1,2,\ldots,n) \qquad (4.3.35)$$

$$\frac{\partial^2 \log L}{\partial\Delta\partial\bar{\beta}_k} = - \frac{1}{2} \sum_{i=1}^{n} [\Delta + \sigma_{ii}(X_i'X_i)^{-1}]^{-1}[\underline{i}_k(\underline{b}_i - \underline{\bar{\beta}})' + (\underline{b}_i - \underline{\bar{\beta}})\underline{i}_k']$$

$$\cdot [\Delta + \sigma_{ii}(X_i'X_i)^{-1}]^{-1} \qquad (k=1,2,\ldots,K) \qquad (4.3.36)$$

where $\bar{\beta}_k$ is the k-th element of $\underline{\bar{\beta}}$ and $\underline{i}_k$ is the k-th column of an identity matrix of order K, cf. Tracy and Dwyer [1969, p. 1580, (3.15)].

Since $E\underline{b}_i = \underline{\bar{\beta}}$, expectations of $\frac{\partial^2 \log L}{\partial\sigma_{ii}\partial\underline{\bar{\beta}}}$ and $\frac{\partial^2 \log L}{\partial\Delta\partial\bar{\beta}_k}$ are equal to zero for every k. Consequently, the information matrix is given by

$$\begin{bmatrix} - E\,\dfrac{\partial^2 \log L}{\partial\bar{\beta}_k\partial\bar{\beta}_{k'}} & - E\,\dfrac{\partial^2 \log L}{\partial\bar{\beta}_k\partial\theta_i} \\ & - E\,\dfrac{\partial^2 \log L}{\partial\theta_i\partial\theta_j} \end{bmatrix} = \begin{bmatrix} \sum_{i=1}^{n} [\Delta + \sigma_{ii}(X_i'X_i)^{-1}]^{-1} & 0 \\ & - E\,\dfrac{\partial^2 \log L}{\partial\theta_i\partial\theta_j} \end{bmatrix}$$

$$(4.3.37)$$

where θ_i is an element of $\underline{\theta}$. Let

$$I(\underline{\bar{\beta}}) = \sum_{i=1}^{n} [\Delta + \sigma_{ii}(X_i'X_i)^{-1}]^{-1} \quad . \qquad (4.3.38)$$

It follows from (4.3.37) that the submatrix of the inverse of information matrix corresponding to $\bar{\beta}$ is equal to (4.3.38). That is, $V[\bar{b}(\underline{\theta})] = [I(\bar{\beta})]^{-1}$ which is called Cramer-Rao lower bound, cf. Rao (1965a, p. 266).

(f) Construction of Forecast Intervals

(1) Under Assumption 4.3.1, $\frac{(T-K)s_{ii}}{\sigma_{ii}} = \frac{\underline{u}_i' M_i \underline{u}_i}{\sigma_{ii}}$ is χ^2 distributed with (T-K) d.f. because M_i is an idempotent matrix of rank (T-K), and $\underline{u}_i$ is a vector of T independently normally distributed variables with mean zero and constant variance σ_{ii}. The quadratic forms $\underline{u}_i' M_i \underline{u}_i$ (i=1,2,...,n) are mutually independent.

(2) For each i, the prediction error $(\underline{b}_i - \underline{\beta}_i) = (X_i' X_i)^{-1} X_i' \underline{u}_i$ and the error sum of squares $\underline{y}_i' M_i \underline{y}_i = \underline{u}_i' M_i \underline{u}_i$ are independently distributed because $(X_i' X_i)^{-1} X_i' M_i = 0$.

(3) It follows from the above argument that

$$\frac{\underline{\ell}' (\underline{b}_i - \underline{\beta}_i)}{[s_{ii} \underline{\ell}' (X_i' X_i)^{-1} \underline{\ell}]^{1/2}} \sim t_{(T-K)} \qquad (4.3.39)$$

for i=1,2,...,n

where $\underset{(1xK)}{\underline{\ell}'}$ is a vector of arbitrary elements.

(4) A forecast interval for $\underline{\ell}' \underline{\beta}_i$ with $(1-\alpha)$ probability is given by

$$\underline{\ell}' \underline{b}_i + [s_{ii} \underline{\ell}' (X_i' X_i)^{-1} \underline{\ell}]^{1/2} t_{(1/2)\alpha} \qquad (4.3.40)$$

for i=1,2,...,n

where $t_{(1/2)\alpha}$ is the upper $(1/2)\alpha$ point of the t-distribution with (T-K) d.f.

The interval statement (4.3.40) indicates the reliability of the forecast $\underline{\ell}' \underline{b}_i$ in predicting $\underline{\ell}' \underline{\beta}_i$ as a linear function of $\underline{y}_i$. The interval in (4.3.40) is similar to a beta-expectation tolerance interval [cf. Fraser and Guttman (1959)] and it can be interpreted as meaning that there is a $1-\alpha$ level of confidence for one observation $\underline{\ell}' \underline{b}_i$ to be covered by the interval (4.3.40).

(5) If we set up the forecast interval (4.3.40) for all the elements of $\underline{\beta}_i$, or for many linear combinations, $\underline{\ell}' \underline{\beta}_i$, we will get as many intervals covering their respective elements of $\underline{\beta}_i$ or linear combinations, $\underline{\ell}' \underline{\beta}_i$. But these say little about the probability of a forecast interval simultaneously covering all the

elements of $\underline{\beta}_i$ or the linear combinations $\underline{\ell}'\underline{\beta}_i$. For this purpose we have to construct simultaneous forecast regions. First recall the familiar formula for the probability that the elements of an observational vector on $\underline{\beta}_i$ are contained in an ellipsoid in the K-dimensional $\underline{\beta}_i$ space. That is,

$$\Pr\left\{\frac{(\underline{b}_i-\underline{\beta}_i)'(X_i'X_i)(\underline{b}_i-\underline{\beta}_i)}{Ks_{ii}} \le F_\alpha\right\} = 1 - \alpha \tag{4.3.41}$$

where F_α is the upper α point of F distribution with K,T-K d.f.

Using the Cauchy-Schwartz inequality we can derive from (4.3.41) the following simultaneous forecast interval which covers all linear combinations of $\underline{\beta}_i$ with probability 1-α.

$$\Pr(\underline{\ell}'\underline{\beta}_i \epsilon \underline{\ell}'\underline{b}_i \pm \left\{s_{ii}KF_\alpha \underline{\ell}'(X_i'X_i)^{-1}\underline{\ell}\right\}^{1/2} \text{ for all } \underline{\ell}) = 1-\alpha \ . \tag{4.3.42}$$

For any particular function $\underline{\ell}'\underline{\beta}_i$,

$$\Pr(\underline{\ell}'\underline{\beta}_i \epsilon \underline{\ell}'\underline{b}_i \pm \left\{s_{ii}KF_\alpha \underline{\ell}'(X_i'X_i)^{-1}\underline{\ell}\right\}^{1/2} \ge 1-\alpha \ , \tag{4.3.43}$$

cf. Rao (1965a, p. 198).

(g) Asymptotic Properties of the Estimator $\bar{\underline{b}}(\underline{\theta})$

We study the consistency of $\bar{\underline{b}}(\underline{\theta})$ under Assumptions 4.2.1, 4.3.1 and the following assumption.[18]

Assumption 4.3.2: $\lim_{T\to\infty} T^{-1}X_i'X_i$ is finite and positive definite. Consider $\underline{b}_i = \underline{\beta}_i + (X_i'X_i)^{-1}X_i'\underline{u}_i$ shown in (4.3.10). We have

$$\underset{T\to\infty}{\text{plim}}|\underline{b}_i-\underline{\beta}_i| = \lim_{T\to\infty}\left[\frac{X_i'X_i}{T}\right]^{-1} \underset{T\to\infty}{\text{plim}} \frac{X_i'\underline{u}_i}{T} = 0 \ . \tag{4.3.44}$$

Following the same argument as in Goldberger (1964, pp. 269-72) we can easily show that

$$\underset{T\to\infty}{\text{plim}} \ s_{ii} = \sigma_{ii} \ . \tag{4.3.45}$$

---

[18] The results that follow could be reproduced without the normality assumption. It would then be necessary to impose certain regularity conditions on the density function of $\underline{y}$.

Taking probability limits on both sides of (4.3.24), we have

$$\operatorname*{plim}_{T\to\infty}\left|\hat{\Delta} - \frac{S_{\underline{\beta}}}{n-1}\right| = \operatorname*{plim}_{T\to\infty}\left|\frac{S_b}{n-1} - \frac{S_{\underline{\beta}}}{n-1}\right| - \frac{1}{n}\sum_{i=1}^{n} \operatorname*{plim}_{T\to\infty} s_{ii}\cdot T^{-1} \operatorname*{plim}_{T\to\infty}\left(X_i'X_iT^{-1}\right)^{-1} \tag{4.3.46}$$

where $S_{\underline{\beta}} = \sum_{i=1}^{n} \underline{\beta}_i\underline{\beta}_i' - \frac{1}{n}\sum_{i=1}^{n} \underline{\beta}_i \sum_{i=1}^{n} \underline{\beta}_i'$.

The value of the second term on the r.h.s. of (4.3.46) is obviously zero because of Assumption 4.3.2 and the result (4.3.45). We can find the value of the first term by taking probability limits of each term on the r.h.s. of (4.3.22).

$$\begin{aligned}
\operatorname*{plim}_{T\to\infty} |S_b - S_{\underline{\beta}}| = \sum_{i=1}^{n} \Bigg[& \underline{\beta}_i \cdot \operatorname{plim} \frac{\underline{u}_i'X_i}{T} \cdot \operatorname{plim}\left(\frac{X_i'X_i}{T}\right)^{-1} \\
& + \operatorname{plim}\left(\frac{X_i'X_i}{T}\right)^{-1} \cdot \operatorname{plim}\left(\frac{X_i'\underline{u}_i}{T}\right) \cdot \underline{\beta}_i' \\
& + T^{-1} \lim\left(\frac{X_i'X_i}{T}\right)^{-1} \cdot \operatorname{plim} \frac{X_i'\underline{u}_i\underline{u}_i'X_i}{T} \\
& \cdot \lim\left(\frac{X_i'X_i}{T}\right)^{-1}\Bigg] - \frac{1}{n}\Bigg[\sum_{i=1}^{n} \underline{\beta}_i \sum_{i=1}^{n} \operatorname{plim} \frac{\underline{u}_i'X_i}{T} \\
& \cdot \lim\left(\frac{X_i'X_i}{T}\right)^{-1} + \sum_{i=1}^{n} \lim\left(\frac{X_i'X_i}{T}\right)^{-1} \\
& \cdot \operatorname{plim} \frac{X_i'\underline{u}_i}{T} \cdot \sum_{i=1}^{n} \underline{\beta}_i' + T^{-1} \sum_{i=1}^{n} \lim\left(\frac{X_i'X_i}{T}\right)^{-1} \\
& \cdot \operatorname{plim} \frac{X_i'\underline{u}_i}{T} \cdot \sum_{i=1}^{n} \underline{u}_i'X_i \cdot \operatorname{plim}\left(\frac{X_i'X_i}{T}\right)^{-1}\Bigg].
\end{aligned} \tag{4.3.47}$$

Since each term on the r.h.s. of (4.3.47) is zero,

$$\operatorname*{plim}_{T\to\infty} |S_b - S_{\underline{\beta}}| = 0 \quad ; \tag{4.3.48}$$

cf. Rao (1965a, p. 319).

It is easy to recognize that $S_{\underline{\beta}}$ is the sample variance-covariance matrix of $\underline{\beta}_i$. As $T\to\infty$, the distribution of the sample variance-covariance matrix of $\underline{b}_i$

tends in probability to that of the sample variance-covariance matrix of $\underline{\beta}_i$. For sufficiently large T, with n fixed, we can treat $\underline{b}_i$ (i=1,2,...,n), as a sample of size n from the population of $\underline{\beta}_i$.

Substituting (4.3.48) in (4.3.46) we get

$$\underset{T\to\infty}{\text{plim}}\left|\hat{\Delta} - \frac{S_\beta}{n-1}\right| = 0 . \tag{4.3.49}$$

Considering $\bar{\underline{b}}(\underline{\theta})$ in (4.3.6) we have

$$\underset{T\to\infty}{\text{plim}}\ \bar{\underline{b}}(\underline{\theta}) = \left[\sum_{j=1}^{n}\left\{\text{plim}\ \Delta + \text{plim}\ \frac{\sigma_{jj}}{T}\cdot\text{plim}\left(\frac{X_j'X_j}{T}\right)^{-1}\right\}^{-1}\right]^{-1}$$

$$\cdot\sum_{i=1}^{n}\left\{\text{plim}\ \Delta + \text{plim}\ \frac{\sigma_{ii}}{T}\cdot\text{plim}\left(\frac{X_i'X_i}{T}\right)^{-1}\right\}^{-1}\left(\underline{\beta}_i + \text{plim}\left(\frac{X_i'X_i}{T}\right)^{-1}\cdot\text{plim}\ \frac{X_i'u_i}{T}\right) ,$$

$$\text{i.e.,}\quad \underset{T\to\infty}{\text{plim}}\left|\bar{\underline{b}}(\underline{\theta}) - \underline{m}_\beta\right| = 0 \tag{4.3.50}$$

where $\underline{m}_\beta = \frac{1}{n}\sum_{i=1}^{n}\underline{\beta}_i$. Thus, the mean of the asymptotic distribution of $\bar{\underline{b}}(\underline{\theta})$ is equal to the mean of $\underline{\beta}_i$. Invoking the law of large numbers across individuals we can say that $\underset{n\to\infty}{\text{plim}}\ \underline{m}_\beta = \bar{\underline{\beta}}$ because $\bar{\underline{\beta}}$ is assumed to be finite. Therefore, $\bar{\underline{b}}(\underline{\theta})$ is a consistent estimator of $\bar{\underline{\beta}}$. Since Δ is assumed to be finite, $\frac{S_\beta}{n-1}$ converges in probability to Δ as $n\to\infty$.

In view of results in (4.3.44), (4.3.45) and (4.3.49) it follows from Slutsky's theorem [cf. Goldberger (1966, p. 119)] that

$$\underset{T\to\infty}{\text{plim}}\left|\bar{\underline{b}}(\hat{\underline{\theta}}) - \underline{m}_\beta\right| = 0 . \tag{4.3.51}$$

Since $\underset{n\to\infty}{\text{plim}}\ \underline{m}_\beta = \bar{\underline{\beta}}$, $\underline{b}(\hat{\underline{\theta}})$ is also a consistent estimator of $\bar{\underline{\beta}}$.

From (4.3.51) it follows that the asymptotic distribution of $\underline{b}(\hat{\underline{\theta}})$ is the same as the distribution of $\underline{m}_\beta$ [cf. Rao (1965a, p. 101)]. Since $\frac{\Delta}{n}$ is the variance-covariance matrix of $\underline{m}_\beta$, we treat $\frac{\hat{\Delta}}{n}$ as the estimated asymptotic variance-covariance matrix of $\underline{b}(\hat{\underline{\theta}})$.

(h) Asymptotic Efficiency of $\underline{\bar{b}}(\hat{\underline{\theta}})$

The consistent estimator $\underline{\bar{b}}(\hat{\underline{\theta}})$ is said to be efficient if

$$\sqrt{n}\left| \underline{\bar{b}}(\hat{\underline{\theta}}) - \underline{\bar{\beta}} - [n^{-1}I(\underline{\bar{\beta}})]^{-1} \frac{1}{n} \frac{\partial \log p(\underline{y}|\underline{\bar{\beta}},\underline{\theta})}{\partial \underline{\bar{\beta}}} \right| \to 0 \qquad (4.3.52)$$

in probability (i.p.) or with probability 1, where $p(\underline{y}|\underline{\bar{\beta}},\underline{\theta})$ is the joint probability density function (pdf) of the random vectors $\underline{y}_1,\underline{y}_2,\ldots,\underline{y}_n$, cf. Rao (1965a, p. 285). If (4.3.52) is true, then $\underline{\bar{b}}(\hat{\underline{\theta}})$ is said to have the first order efficiency, cf. Rao (1963).

Combining (4.3.29) and (4.3.38) we have

$$[n^{-1}I(\underline{\bar{\beta}})]^{-1} \frac{1}{n} \frac{\partial \log p(\underline{y}|\underline{\bar{\beta}},\underline{\theta})}{\partial \underline{\bar{\beta}}} = \underline{\bar{b}}(\underline{\theta}) - \underline{\bar{\beta}} \quad . \qquad (4.3.53)$$

Consequently, under Assumption 4.3.1 the condition (4.3.52) can be written as

$$\sqrt{\underline{n}}\left|[\underline{\bar{b}}(\hat{\underline{\theta}}) - \underline{\bar{\beta}}] - [\underline{\bar{b}}(\underline{\theta}) - \underline{\bar{\beta}}]\right| \to 0 \qquad (4.3.54)$$

i.p. as $n\to\infty$, $T\to\infty$. The results (4.3.50) and (4.3.51) clearly indicate that $\underline{\bar{b}}(\hat{\underline{\theta}})$ satisfies the condition (4.3.54). Using the multivariate central limit theorem given in Rao [1965a, p. 108, Theorem (iv)], we can easily show that the asymptotic distribution of

$$\frac{1}{\sqrt{n}} \frac{\partial \log p(\underline{y}|\underline{\bar{\beta}},\underline{\theta})}{\partial \underline{\bar{\beta}}} \qquad (4.3.55)$$

is K-variate normal with mean zero and variance-covariance matrix equal to $\lim_{\substack{n\to\infty \\ T\to\infty}} n^{-1}I(\underline{\bar{\beta}})$. Consequently, the condition under Assumption 4.3.1 implies that $\sqrt{n}[\underline{\bar{b}}(\hat{\underline{\theta}}) - \underline{\bar{\beta}}]$ is also asymptotically K-variate normal with mean 0 and the variance-covariance matrix equal to $\lim_{\substack{n\to\infty \\ T\to\infty}} [n^{-1}I(\underline{\bar{\beta}})]^{-1}$ which is the limit of the inverse of information matrix. Hence, $\underline{\bar{b}}(\hat{\underline{\theta}})$ is asymptotically efficient.

(i) Testing of Hypotheses on the Means of Coefficients

It follows from (4.3.49) and (4.3.51) that

$$\left| \frac{\underline{\ell}' \sqrt{n}(\underline{\bar{b}}(\hat{\theta})-\underline{\bar{\beta}})}{[\underline{\ell}'\hat{\Delta}\underline{\ell}]^{1/2}} - \frac{\underline{\ell}' \sqrt{n}(\underline{m}_\beta-\underline{\bar{\beta}})}{[\frac{1}{n-1}\underline{\ell}'S_\beta\underline{\ell}]^{1/2}} \right| \to 0 \quad \text{i.p.} \tag{4.3.56}$$

as $T\to\infty$ and n is fixed. According to the limit theorem (ix) in Rao (1965a, p. 101), the asymptotic distribution (as $T\to\infty$ with n fixed) of $\underline{\ell}' \sqrt{n}[\underline{\bar{b}}(\hat{\theta})-\underline{\bar{\beta}}]/[\underline{\ell}'\hat{\Delta}\underline{\ell}]^{1/2}$ is the same as the distribution of $\underline{\ell}' \sqrt{n}(\underline{m}_\beta-\underline{\bar{\beta}})/[\frac{1}{n-1}\underline{\ell}'S_\beta\underline{\ell}]^{1/2}$.

Under Assumptions 4.2.1 and 4.3.1 the matrix S_β is Wishart distributed with n-1 d.f., and $E\frac{S_\beta}{n-1} = \Delta$. The distribution of $\underline{m}_\beta$ is normal with mean $\underline{\bar{\beta}}$ and variance-covariance matrix $\frac{\Delta}{n}$. Furthermore, $\underline{m}_\beta$ which is the sample mean, is independent of $\frac{S_\beta}{n-1}$ which is the sample variance-covariance matrix, cf. Rao (1965a, p. 454). It can also be derived from a well known result that a linear combination of the elements of the Wishart matrix, $\underline{\ell}'S_\beta\underline{\ell}$ is $\chi^2\sigma_\ell^2$ distributed with n-1 d.f., where $\sigma_\ell^2 = \underline{\ell}'\Delta\underline{\ell}$, cf. Rao (1965a, p. 452). Therefore the distribution of $\underline{\ell}' \sqrt{n}(\underline{m}_\beta-\underline{\bar{\beta}})/[\frac{1}{n-1}\underline{\ell}'S_\beta\underline{\ell}]^{1/2}$ is Student's t with n-1 d.f. and the asymptotic distribution of

$$\frac{\underline{\ell}' \sqrt{n}(\underline{\bar{b}}(\hat{\theta}) - \underline{\bar{\beta}})}{[\underline{\ell}'\hat{\Delta}\underline{\ell}]^{1/2}} \tag{4.3.57}$$

is Student's t with n-1 d.f. We can use the statistic (4.3.57) for testing the hypotheses on $\underline{\bar{\beta}}$. The power functions of such tests in small samples are not known since the exact finite sample distribution of the statistics (4.3.57) is not known. However, asymptotically the test of a hypothesis on $\underline{\bar{\beta}}$ against a class of alternative hypotheses using (4.3.57) is equivalent to the Student's t-test of the same hypothesis against the same class of alternatives because of (4.3.56). Thus, on the basis of the optimal properties of Student's t-test we can give asymptotic justification for using (4.3.57) as a test-statistic.

From an asymptotic test-statistic we are not usually justified in constructing a $(1-\alpha)$ confidence interval for a parameter unless the convergence to a limiting distribution is uniform in compact intervals of the parameter, cf. Rao

(1965a, p. 287). Since $\bar{\underline{b}}(\hat{\underline{\theta}})$ satisfies the first order efficiency condition (4.3.52), $\bar{\underline{b}}(\hat{\underline{\theta}})$ is a consistent and uniformly asymptotically normal (CUAN) estimator, cf. Rao (1963). For a CUAN estimator we can construct confidence intervals using the asymptotic test-statistic. Thus,

$$\underline{\ell}'\bar{\underline{b}}(\hat{\underline{\theta}}) \pm \frac{t_{\alpha/2}[\underline{\ell}'\hat{\Delta}\underline{\ell}]^{1/2}}{\sqrt{n}} \tag{4.3.58}$$

gives a (1-α) confidence interval for $\underline{\ell}'\bar{\underline{\beta}}$. Here $t_{\alpha/2}$ is the upper $\alpha/2$ point of the t-distribution with n-1 d.f.

(j) Testing of Hypotheses on the Mean Vector of Coefficients

In this section we suggest a criterion for testing the hypothesis, $\bar{\underline{\beta}} = \bar{\underline{\beta}}_0$ (a preassigned value) under Assumptions 4.2.1 and 4.3.1.

Using the results in (4.3.48) and (4.3.51) we can show that

$$|\bar{\underline{b}}(\hat{\underline{\theta}})-\bar{\underline{\beta}})'C(\hat{\underline{\theta}})^{-1}\ (\bar{\underline{b}}(\hat{\underline{\theta}})-\bar{\underline{\beta}})-n\ (n-1)(\underline{m}_\beta-\bar{\underline{\beta}})'S_\beta^{-1}(\underline{m}_\beta-\bar{\underline{\beta}})| \rightarrow 0 \tag{4.3.59}$$

i.p. as $T\rightarrow\infty$ and n is fixed. $C(\hat{\underline{\theta}})$ is defined in (4.3.25). Applying the limit theorem (ix) in Rao (1965a, p. 101) we can show that the asymptotic distribution of the statistic $(\bar{\underline{b}}(\hat{\underline{\theta}})-\bar{\underline{\beta}})'C(\hat{\underline{\theta}})^{-1}\ (\bar{\underline{b}}(\hat{\underline{\theta}})-\bar{\underline{\beta}})$ is the same as the distribution of $n(n-1)(\underline{m}_\beta-\bar{\underline{\beta}})'S_\beta^{-1}(\underline{m}_\beta-\bar{\underline{\beta}})$ which can be easily recognized as Hotelling's T^2-statistic. Rao (1965a, p. 458) presented an elegant derivation (attributable to Wijsman) of Hotelling's distribution. Following exactly the same derivation and adding the result (4.3.59) to it we can show that under the hypothesis $E\underline{\beta}_1 = E\underline{\beta}_2 = \ldots = E\underline{\beta}_n = \bar{\underline{\beta}}_0$ (a preassigned value) the asymptotic distribution of the statistic

$$\frac{(n-K)}{K(n-1)}\ (\bar{\underline{b}}(\hat{\underline{\theta}})-\bar{\underline{\beta}})'C(\hat{\underline{\theta}})^{-1}\ (\bar{\underline{b}}(\hat{\underline{\theta}})-\bar{\underline{\beta}}) \tag{4.3.60}$$

is F with K and n-K d.f. Since the exact finite sample distribution of the above statistic is not known, the power of a test based on (4.3.60) cannot be discussed here. Since the statistic (4.3.60) differs from T^2-statistic by a quantity which has zero probability limit as $T\rightarrow\infty$, we can say that asymptotically the test of the above hypothesis using (4.3.60) is equivalent to T^2-test of the same hypothesis. The optimal properties of T^2-test are well known, cf. Anderson (1958, pp. 115-118).

(k) Testing of Linear Hypotheses on the Variances of Coefficients

We have shown in subsection 4.3(c) that

$$\hat{\Delta} = \frac{S_b}{n-1} - \frac{1}{n} \sum_{i=1}^{n} s_{ii}(X_i'X_i)^{-1}$$

is an unbiased estimator of Δ. It follows from (4.3.49) that $|\underline{\ell}'\hat{\Delta}\underline{\ell} - \underline{\ell}' \frac{S_\beta}{n-1}\underline{\ell}| \to 0$ in probability as $T\to\infty$ and n is fixed. Since $S_\beta \sim W_k\ (n-1,\Delta)$, $\underline{\ell}'S_\beta\underline{\ell}$ is $\chi^2\sigma_\ell^2$ distributed with n-1 d.f. σ_ℓ^2 is equal to $\underline{\ell}'\Delta\underline{\ell}$. Therefore the asymptotic distribution of $(n-1)\underline{\ell}'\hat{\Delta}\underline{\ell}$ is $\chi^2\sigma_\ell^2$ with n-1 d.f. The statistic

$$(n-1)\ \frac{\underline{\ell}'\hat{\Delta}\underline{\ell}}{\underline{\ell}'\hat{\Delta}_0\underline{\ell}} \qquad (4.3.61)$$

can be used to test the hypothesis that $\Delta = \Delta_0$, not all elements of which are zero. Again we can only give asymptotic justification for the use of this test.

(l) A Test for the Randomness of Coefficients

In order to investigate whether the specification of a random coefficient model is appropriate to a given situation one would like to test whether the coefficients in a model are indeed random. This can be accomplished by testing the hypothesis that the variance-covariance matrix of a coefficient vector is null. We develop in this section a test for this hypothesis following the likelihood-ratio (LR) criterion.

Let the null hypothesis be,

$$H_0:\ \Delta = 0 \text{ given that the } E\underline{\beta}_i = \bar{\underline{\beta}} \text{ for all i.} \qquad (4.3.62)$$

The LR criterion to test H_0 is

$$\Lambda_c = \sup_{H_0} L(\bar{\underline{\beta}},\underline{\theta}|\underline{y},X) / \sup_{\bar{\underline{\beta}},\underline{\theta}} L(\bar{\underline{\beta}},\underline{\theta}|\underline{y},X)\ . \qquad (4.3.63)$$

Maximizing the likelihood (4.3.28) under the restriction $\Delta = 0$, we have

$$\underline{\hat{\beta}}(\underline{\sigma}) = \left[\sum_{j=1}^{n} \frac{X_j'X_j}{\sigma_{jj}}\right]^{-1} \left[\sum_{i=1}^{n} \frac{X_i'\underline{y}_i}{\sigma_{ii}}\right] \quad (4.3.64)$$

and

$$\hat{\sigma}_{ii} = [\underline{y}_i - X_i\underline{\hat{\beta}}(\underline{\sigma})]'[\underline{y}_i - X_i\underline{\hat{\beta}}(\underline{\sigma})]/T \quad (4.3.65)$$

where $\underline{\sigma} \equiv (\sigma_{11},\ldots,\sigma_{nn})'$. Taking a consistent estimate of σ_{ii}, say s_{ii} in (4.3.21), as a starting value for σ_{ii} we can solve (4.3.64) and (4.3.65) by an iterative procedure until we obtain stable estimates for $\underline{\overline{\beta}}$ and $\underline{\sigma}$. Substituting these in (4.3.28) we have

$$\sup_{\Delta=0} L(\underline{\overline{\beta}},\underline{\theta}|\underline{y},X) = (2\pi)^{-\frac{nT}{2}} \prod_{i=1}^{n} \hat{\sigma}_{ii}^{-T/2} e^{-\frac{nT}{2}} . \quad (4.3.66)$$

ML estimates of $\underline{\overline{\beta}}$ and $\underline{\theta}$ when $\Delta \neq 0$, are already derived in subsection 4.3(d). Substituting these in (4.3.28) we have

$$\sup_{\underline{\overline{\beta}},\underline{\theta}} L(\underline{\overline{\beta}},\underline{\theta}|\underline{y},X) = L(\underline{\overline{b}}(\underline{\tilde{\theta}}),\underline{\tilde{\theta}}|\underline{y},X) \quad . \quad (4.3.67)$$

Taking the ratio of (4.3.66) to (4.3.67) we obtain $\hat{\Lambda}_c$. The densities of $\underline{y}_i$ satisfy the conditions underlying the multivariate central limit theorem for nonidentically distributed random variables stated in Rao (1965a, p. 118). Following Silvey (1964) and Hartley and Rao (1967) we can show that under certain general assumptions $\underline{\overline{b}}(\underline{\tilde{\theta}})$ and $\underline{\tilde{\theta}}$ are asymptotically normal and efficient. Consequently, $-2 \log_e \hat{\Lambda}_c$ is asymptotically distributed, under the null hypothesis, as a central χ^2 with $\frac{1}{2}$ K(K+1)d.f. At the present stage of our work we can only rely on the asymptotic properties of the test based on $\hat{\Lambda}_c$ because the derivation of either the finite sample distribution of $\hat{\Lambda}_c$ or the exact moments of $\hat{\Lambda}_c$ is complicated. Therefore the power of the test based on $\hat{\Lambda}_c$ in small samples is unknown. In order to avoid the complicated steepest ascent method we suggest an alternative approach which is asymptotically equivalent to the above procedure. Under Assumption 4.3.2 we have shown in (4.3.45) and (4.3.46) that s_{ii} and $n^{-1}S_b$ are consistent estimators of σ_{ii} and Δ respectively. Notice that s_{ii} and $n^{-1}S_b$ are the solutions of (4.3.30) and (4.3.31) respectively when T is sufficiently large because $\lim_{T\to\infty} T^{-1}\text{tr}\Delta^{-1}$ $\cdot(T^{-1}X_i'X_i)^{-1} = 0$ and $\lim_{T\to\infty} [\Delta + T^{-1}\sigma_{ii}(T^{-1}X_i'X_i)^{-1}] = \Delta$. Substituting s_{ii} and $n^{-1}S_b$ for σ_{ii} and $\Delta + \sigma_{ii}(X_i'X_i)^{-1}$ respectively in (4.3.28) we have

$$(2\pi)^{-\frac{nT}{2}} \prod_{i=1}^{n} \left\{ s_{ii}^{-\frac{1}{2}(T-K)} |X_i'X_i|^{-\frac{1}{2}} |n^{-1}S_b|^{-\frac{1}{2}} \right\} e^{-\frac{nT}{2}} \quad . \tag{4.3.68}$$

The difference between (4.3.67) and (4.3.68) will be negligible if T is large. Since we are developing an asymptotic test of the hypothesis (4.3.62), we can ignore this difference. If we take the ratio of (4.3.66) to (4.3.68) we have

$$\hat{\Lambda}_c^* = \frac{\prod_{i=1}^{n} \hat{\sigma}_{ii}^{-\frac{T}{2}}}{\prod_{i=1}^{n} \left\{ s_{ii}^{-\frac{1}{2}(T-K)} |X_i'X_i|^{-\frac{1}{2}} |n^{-1}S_b|^{-\frac{1}{2}} \right\}} \cdot \tag{4.3.69}$$

Since $\underset{T\to\infty}{\text{plim}}|\tilde{s}_{ii}-s_{ii}| = 0$ and $\underset{T\to\infty}{\text{plim}}|\tilde{\Delta}-n^{-1}S_b| = 0$, we can show that $\text{plim}|-2\log_e\hat{\Lambda}_c + 2\log_e\Lambda_c^*| = 0$. Consequently, it follows from a limit theorem in Rao (1965a, p. 101) that the asymptotic distribution of $-2\log_e\hat{\Lambda}_c^*$ can be approximated by central χ^2 with $\frac{1}{2}K(K+1)$d.f. when the null hypothesis (4.3.62) is true. We can use the asymptotic distribution of

$$-2\log_e\hat{\Lambda}_c^* = T\sum_{i=1}^{n}\log_e\hat{\sigma}_{ii} - (T-K)\sum_{i=1}^{n}\log_e s_{ii} - \sum_{i=1}^{n}\log_e|X_i'X_i| - n\log_e|n^{-1}S_b| \tag{4.3.70}$$

to set up a critical region for the test of the hypothesis (4.3.62). This test procedure is useful in verifying whether (4.3.16) is close to a null matrix.

(m) Asymptotic Test of Equality Between Fixed Coefficient Vectors in n Relations

Before attempting to estimate a model of the type (4.2.1) it is better to test whether the fixed coefficient vectors $\underline{\beta}_1,\ldots,\underline{\beta}_n$ are equal to one another. Depending upon the outcome of this preliminary test we can decide whether Assumption 4.2.1 is tenable for the situation under study. Consider the following null hypothesis.

$$H_0 : \underline{\beta}_1 = \underline{\beta}_2 = \dots = \underline{\beta}_n = \underline{\beta} \text{ given that } \Delta = 0. \tag{4.3.71}$$

An asymptotic procedure of testing the hypothesis (4.3.71) is developed by Zellner (1962a) and we only sketch his proof here. The likelihood function of $\underline{y}$ under the hypothesis (4.3.71) is

$$L(\underline{\beta},\underline{\sigma}|\underline{y},X) = (2\pi)^{-\frac{nT}{2}} \prod_{i=1}^{n} \sigma_{ii}^{-T/2} \cdot \exp\left\{-\frac{1}{2}\left[\sum_{i=1}^{n} (\underline{y}_i - X_i\underline{\beta})'(\underline{y}_i - X_i\underline{\beta})/\sigma_{ii}\right]\right\} . \tag{4.3.72}$$

Maximizing (4.3.72) with respect to $\underline{\beta}$ and $\underline{\sigma}$ we obtain (4.3.64) and (4.3.65) as the estimators of $\underline{\beta}$ and σ_{ii} respectively. Substituting these in (4.3.72) we have

$$\sup_{H_0} L(\underline{\beta},\underline{\sigma}|\underline{y},X) = (2\pi)^{-\frac{nT}{2}} \prod_{i=1}^{n} \hat{\sigma}_{ii}^{-\frac{T}{2}} e^{-\frac{nT}{2}} . \tag{4.3.73}$$

The likelihood function under the alternative hypothesis, $H_1 : \underline{\beta}_1 \neq \underline{\beta}_2 \neq \dots \neq \underline{\beta}_n$, is

$$L(\underline{\beta}_1,\dots,\underline{\beta}_n\underline{\sigma}|\underline{y},X) = (2\pi)^{-\frac{nT}{2}} \prod_{i=1}^{n} \sigma_{ii}^{-\frac{T}{2}} \cdot \exp\left\{-\frac{1}{2}\left[\sum_{i=1}^{n} (\underline{y}_i - X_i\underline{\beta}_i)'(\underline{y}_i - X_i\underline{\beta}_i)/\sigma_{ii}\right]\right\} \tag{4.3.74}$$

which upon substitution of ML estimates of $\underline{\beta}_1,\dots,\underline{\beta}_n$ and $\underline{\sigma}$ becomes

$$\sup_{H_1} L(\underline{\beta}_1,\dots,\underline{\beta}_n,\underline{\sigma}|\underline{y},X) = (2\pi)^{-\frac{nT}{2}} \prod_{i=1}^{n} [(T-K)s_{ii}/T]^{-T/2} e^{-\frac{nT}{2}} . \tag{4.3.75}$$

The LR criterion is

$$\hat{\lambda}_c = \left\{\prod_{i=1}^{n} \hat{\sigma}_{ii}^{-T/2}\right\} \left\{\prod_{i=1}^{n} [(T-K)s_{ii}/T]^{T/2}\right\} \tag{4.3.76}$$

and

$$-2 \log_e \hat{\lambda}_c = T \sum_{i=1}^{n} \log_e \left[\frac{T\hat{\sigma}_{ii}}{(T-K)s_{ii}}\right] \tag{4.3.77}$$

which, under the hypothesis (4.3.71) is asymptotically distributed as central χ^2 with (n-1)K d.f. as $T\to\infty$ and n is fixed. We can write

$$-2 \log_e \hat{\lambda}_c = T \sum_{i=1}^{n} \log_e \left[1 + \frac{(\underline{b}_i - \hat{\underline{\beta}}(\underline{\sigma}))' X_i' X_i (\underline{b}_i - \hat{\underline{\beta}}(\underline{\sigma}))}{(T-K)s_{ii}}\right] \quad . \tag{4.3.78}$$

By expanding the r.h.s. of (4.3.78) into an infinite series we have

$$-2 \log_e \hat{\lambda}_c = \sum_{i=1}^{n} \frac{(\underline{b}_i - \hat{\underline{\beta}}(\underline{\sigma}))' X_i' X_i (\underline{b}_i - \hat{\underline{\beta}}(\underline{\sigma}))}{s_{ii}} + o_p(T^{-1}) \quad . \tag{4.3.79}$$

Consequently, according to a limit theorem in Rao (1965a, p. 101), both $-2 \log_e \hat{\lambda}_c$ and the first term on the r.h.s. of (4.3.79) have the same limiting distribution. Since the statistic, $\sum_{i=1}^{n} \underline{y}_i'\underline{y}_i/\sigma_{ii} - \sum_{i=1}^{n} \underline{y}_i' X_i \underline{b}_i/\sigma_{ii}$ is χ^2 distributed with n(T-K)d.f. and s_{ii} is a consistent estimator of σ_{ii}, the statistic $\left[\sum_{i=1}^{n} \underline{y}_i'\underline{y}_i/s_{ii} - \sum_{i=1}^{n} \underline{y}_i' X_i \underline{b}_i/s_{ii}\right]$ is also asymptotically χ^2 with n(T-K) d.f. But $\left[\sum_{i=1}^{n} \underline{y}_i'\underline{y}_i/s_{ii} - \sum_{i=1}^{n} \underline{y}_i' X_i \underline{b}_i/s_{ii}\right] =$ n(T-K). It is easy to verify that the statistic $\left[\sum_{i=1}^{n} \underline{y}_i'\underline{y}_i/\sigma_{ii} - \sum_{i=1}^{n} \underline{y}_i' X_i \underline{b}_i/\sigma_{ii}\right]$ is independent of the statistic $\sum_{i=1}^{n} \frac{(\underline{b}_i - \hat{\underline{\beta}}(\underline{\sigma}))' X_i' X_i (\underline{b}_i - \hat{\underline{\beta}}(\underline{\sigma}))}{\sigma_{ii}}$. Consequently, Zellner (1962a) is able to show that the asymptotic distribution of

$$\frac{1}{(n-1)K} \sum_{i=1}^{n} \frac{(\underline{b}_i - \underline{\beta}(\underline{\sigma}))' X_i' X_i (\underline{b}_i - \underline{\beta}(\underline{\sigma}))}{s_{ii}} \tag{4.3.80}$$

can be well approximated by F-distribution with (n-1)K, n(T-K) d.f. for large T. Zellner (1962a) placed the above test procedure in a more general regression setting. We can use the asymptotic distribution of (4.3.80) for testing the hypothesis (4.3.71).

4.4 Estimation of Parameters in RCR Models when Disturbances are Serially Correlated

In Section 4.3 we considered the estimation of parameters in a RCR model when disturbances are not serially correlated. It is possible that the disturbances in (4.2.1) may exhibit serial correlation. In particular, when we combine time series with cross-section data, the micro-units' disturbance terms may display an autoregressive structure.

We analyze the system shown in (4.2.1) under the following assumption. Let $\underline{u}_i \equiv (u_{i1},\ldots,u_{it},\ldots,u_{iT})'$.

Assumption 4.4.1: $u_{it} = \rho_i u_{it-1} + \epsilon_{it}$ $(0<|\rho_i|<1)$ and $E\epsilon_{it} = 0$, $E\epsilon_{it}\epsilon_{jt'} = \sigma_{ii}$ if $t = t'$ and $i = j$, 0 otherwise. The ρ_i $(i=1,2,\ldots,n)$ are fixed parameters. The specifications (1), (2), (4) and (5) in Assumption 4.2.1 remain as they are.

Under the above assumption the variance-covariance matrix of $\underline{u}_i$'s in (4.2.1) takes the following form. For $i = j$, $E\underline{u}_i\underline{u}_j'$ is equal to

$$\sigma_{ii}\Omega_{ii} = \frac{\sigma_{ii}}{1-\rho_i^2}\begin{bmatrix} 1 & \rho_i & \rho_i^2 & \cdots & \rho_i^{T-1} \\ \rho_i & 1 & \rho_i & \cdots & \rho_i^{T-2} \\ \cdot & \cdot & \cdot & & \cdot \\ \cdot & \cdot & \cdot & & \cdot \\ \cdot & \cdot & \cdot & & \cdot \\ \rho_i^{T-1} & \rho_i^{T-2} & \rho_i^{T-3} & \cdots & 1 \end{bmatrix} \tag{4.4.1}$$

and for $i \neq j$, $E\underline{u}_i\underline{u}_j' = 0$.

Furthermore, the variance-covariance matrix of $D(X)\underline{\delta} + \underline{u}$ in (4.3.2a) takes the following form.

$$H(\underline{\Theta}) = \begin{bmatrix} X_1\Delta X_1' + \sigma_{11}\Omega_{11} & 0 & \cdots & 0 \\ 0 & X_2\Delta X_2' + \sigma_{22}\Omega_{22} & \cdots & 0 \\ \cdot & \cdot & & \cdot \\ \cdot & \cdot & & \cdot \\ \cdot & \cdot & & \cdot \\ 0 & 0 & \cdots & X_n\Delta X_n' + \sigma_{nn}\Omega_{nn} \end{bmatrix} \tag{4.4.2}$$

where $H(\underline{\Theta})$ is a symmetric nTxnT matrix, 0's are null matrices of order T, and Θ is a $\frac{1}{2}[K(K+1)+4n]x1$ vector containing all the distinct elements of Δ, $\underline{\sigma}$ and ρ_i $(i=1,2,\ldots,n)$. We assume that each element of $H(\underline{\Theta})$ has continuous first order derivatives and $H(\underline{\Theta})$ is positive definite for all the admissible values of $\underline{\Theta}$. Since the elements of $\underline{\Theta}$ are usually unknown, we develop an Aitken estimator of $\bar{\underline{\beta}}$ based on an estimate of $\underline{\Theta}$.

A consistent estimator of ρ_i is given by

$$\hat{\rho}_i = \frac{\sum_{t=2}^{T} \hat{u}_{it}\hat{u}_{it-1}}{\sum_{t=2}^{T} \hat{u}_{it-1}^2} \tag{4.4.3}$$

where $\hat{u}_{it}$ is the t th element of $\underline{\hat{u}}_i = \underline{y}_i - X_i\underline{b}_i$. Let

$$\hat{R}_i = \begin{bmatrix} \sqrt{1-\hat{\rho}_i^2} & 0 & 0 & \cdots & 0 & 0 \\ -\hat{\rho}_i & 1 & 0 & \cdots & 0 & 0 \\ 0 & -\hat{\rho}_i & 1 & \cdots & 0 & 0 \\ \cdot & \cdot & \cdot & & \cdot & \cdot \\ \cdot & \cdot & \cdot & & \cdot & \cdot \\ \cdot & \cdot & \cdot & & \cdot & \cdot \\ 0 & 0 & 0 & \cdots & -\hat{\rho}_i & 1 \end{bmatrix} \tag{4.4.4}$$

Notice that $\hat{R}_i'\hat{R}_i = \hat{\Omega}_{ii}^{-1}$ where $\hat{\Omega}_{ii}$ is obtained by replacing ρ_i by $\hat{\rho}_i$ in Ω_{ii}.

The variance-covariance matrix of the error term $X_i\underline{\delta}_i + \underline{u}_i$ in the i-th equation is $X_i\Delta X_i' + \sigma_{ii}\Omega_{ii}$. By regressing $\underline{y}_i$ upon X_i and applying Aitken's generlized least squares we obtain

$$\underline{b}_i(\rho_i) = [X_i'(X_i \Delta X_i' + \sigma_{ii}\Omega_{ii})^{-1}X_i]^{-1}[X_i'(X_i\Delta X_i' + \sigma_{ii}\Omega_{ii})^{-1}\underline{y}_i] \quad . \tag{4.4.5}$$

Using a matrix result in Rao (1965a, p. 29, Problem 2.9) we have

$$\begin{aligned}(X_i \Delta X_i' + \sigma_{ii}\Omega_{ii})^{-1} &= \sigma_{ii}^{-1}\Omega_{ii}^{-1} - \sigma_{ii}^{-1}\Omega_{ii}^{-1}X_i\ (X_i'\Omega_{ii}^{-1}X_i)^{-1}X_i'\Omega_{ii}^{-1} \\ &\quad + \Omega_{ii}^{-1}X_i(X_i'\Omega_{ii}^{-1}X_i)^{-1}\left\{(X_i'\Omega_{ii}^{-1}X_i)^{-1}\sigma_{ii} + \Delta\right\}^{-1} \\ &\quad \cdot(X_i'\Omega_{ii}^{-1}X_i)^{-1}X_i'\Omega_{ii}^{-1} \quad .\end{aligned} \tag{4.4.6}$$

Substituting (4.4.6) in (4.4.5) we have

$$\begin{aligned}\underline{b}_i(\rho_i) &= \left\{\Delta + \sigma_{ii}(X_i'\Omega_{ii}^{-1}X_i)^{-1}\right\}\left\{\Delta + \sigma_{ii}(X_i'\Omega_{ii}^{-1}X_i)^{-1}\right\}^{-1} \\ &\quad \cdot(X_i'\Omega_{ii}^{-1}X_i)^{-1}\ X_i'\Omega_{ii}^{-1}\underline{y}_i \\ &= (X_i'\Omega_{ii}^{-1}X_i)^{-1}X_i'\Omega_{ii}^{-1}\underline{y}_i \quad .\end{aligned} \tag{4.4.7}$$

This result is not surprising in view of a theorem in Rao (1967, p. 364, (68)).

Consequently, we can estimate σ_{ii} by

$$s_{ii}^* = \hat{\underline{\varepsilon}}_i'\hat{\underline{\varepsilon}}_i/(T-K) \qquad (i=1,2,\ldots,n) \tag{4.4.8}$$

where $\hat{\underline{\varepsilon}}_i = \hat{R}_i\underline{y}_i - \hat{R}_iX_i\underline{b}_i(\hat{\rho}_i)$ and $\underline{b}_i(\hat{\rho}_i) = (X_i'\hat{R}_i'\hat{R}_iX_i)^{-1}X_i'\hat{R}_i'\hat{R}_i\underline{y}_i$.

We can estimate Δ by

$$\hat{\Delta}(\hat{\rho}) = \frac{S_b(\hat{\rho})}{n-1} - \frac{1}{n}\sum_{i=1}^{n} s_{ii}^*(X_i'\hat{\Omega}_{ii}X_i)^{-1} \tag{4.4.9}$$

where $S_b(\hat{\rho}) = \sum_{i=1}^{n}\underline{b}_i(\hat{\rho}_i)\underline{b}_i'(\hat{\rho}_i) - \frac{1}{n}\sum_{i=1}^{n}\underline{b}_i(\hat{\rho}_i)\sum_{i=1}^{n}\underline{b}_i'(\hat{\rho}_i)$.

An Aitken estimator of $\underline{\bar{\beta}}$ based on the estimators (4.4.3), (4.4.9) and (4.4.8) is

$$\bar{\underline{b}}(\hat{\underline{\Theta}}) = (X'H(\hat{\underline{\Theta}})^{-1}X)^{-1}X'H(\hat{\underline{\Theta}})^{-1}\underline{y}$$

$$= \left[\sum_{j=1}^{n} X_j'(X_j\hat{\Delta}(\hat{\rho})X_j' + s^*_{jj}\hat{\Omega}_{jj})^{-1}X_j\right]^{-1}$$

$$\cdot\left[\sum_{i=1}^{n} X_i'(X_i\hat{\Delta}(\hat{\rho})X_i' + s^*_{ii}\hat{\Omega}_{ii})^{-1}\underline{y}_i\right]$$

$$= \left[\sum_{j=1}^{n} \left\{\hat{\Delta}(\hat{\rho}) + s^*_{jj}(X_j'\hat{\Omega}_{jj}^{-1}X_j)^{-1}\right\}^{-1}\right]^{-1}$$

$$\cdot\left[\sum_{i=1}^{n} \left\{\hat{\Delta}(\hat{\rho}) + s^*_{ii}(X_i'\hat{\Omega}_{ii}^{-1}X_i)^{-1}\right\}^{-1}\underline{b}_i(\hat{\rho}_i)\right] \tag{4.4.10}$$

where use is made of (4.4.6).

Assumption 4.4.2: $\lim_{T\to\infty} T^{-1}X_i'\Omega_{ii}^{-1}X_i$ is finite and positive definite for all i and for $0 < |\rho_i| < 1$.

Following the same argument as in Parks (1967) and utilizing Assumption 4.4.2 we can show that

$$\operatorname*{plim}_{T\to\infty} \hat{\rho}_i = \rho_i, \quad \operatorname*{plim}_{T\to\infty} \hat{\Omega}_{ii} = \Omega_{ii} \text{ and } \operatorname*{plim}_{T\to\infty} s^*_{ii} = \sigma_{ii}\ . \tag{4.4.11}$$

Using Assumption 4.4.2 we can show that the last term on the r.h.s. of (4.4.9) converges in probability to zero and

$$\operatorname*{plim}_{T\to\infty} |\hat{\Delta}(\hat{\rho}) - (n-1)^{-1}S_\beta| = 0\ . \tag{4.4.12}$$

Thus, $\hat{\rho}_i$, s^*_{ii} and $\hat{\Delta}(\hat{\rho})$ are consistent estimators of ρ_i, σ_{ii}, and Δ respectively. Due to the invariance property of consistency $\operatorname*{plim}_{T\to\infty} \hat{\underline{\Theta}} = \underline{\Theta}$. Now

$$\operatorname*{plim}_{T\to\infty} \bar{\underline{b}}(\hat{\underline{\Theta}}) = \operatorname*{plim}_{T\to\infty}\left[\sum_{j=1}^{n} \left\{\hat{\Delta}(\hat{\rho}) + s^*_{jj}(X_j'\hat{\Omega}_{jj}^{-1}X_j)^{-1}\right\}^{-1}\right]^{-1}$$

$$\cdot\left[\sum_{i=1}^{n} \left\{\hat{\Delta}(\hat{\rho}) + s^*_{ii}(X_i'\hat{\Omega}_{ii}^{-1}X_i)^{-1}\right\}^{-1}\underline{b}_i(\hat{\rho}_i)\right]. \tag{4.4.13}$$

It follows from Assumption 4.4.2 and the result (4.4.11) that

$$\operatorname*{plim}_{T\to\infty} \frac{1}{T}\, s^*_{ii}T(X_i'\hat{\Omega}_{ii}^{-1}X_i)^{-1} = 0 \tag{4.4.14}$$

and

$$\underset{T\to\infty}{\text{plim}} \; |\underline{b}_i(\hat{\rho}_i) - \underline{\beta}_i| = 0 \quad . \tag{4.4.15}$$

Substituting (4.4.12), (4.4.14) and (4.4.15) in (4.4.13) we have

$$\underset{T\to\infty}{\text{plim}} \; |\bar{\underline{b}}(\hat{\underline{\Theta}}) - \underline{m}_\beta| = 0 \quad . \tag{4.4.16}$$

This indicates that $\underline{0}$ and Δ are the mean and the variance-covariance matrix respectively of the limiting distribution of $\sqrt{n}[\underline{b}(\hat{\underline{\Theta}})-\bar{\underline{\beta}}]$. When the $\underline{u}_i'$s are autocorrelated, by adopting the estimator $\bar{\underline{b}}(\hat{\underline{\Theta}})$ we will not increase the asymptotic efficiency of an estimator of $\bar{\underline{\beta}}$ because the variance-covariance matrix of the limiting distribution of $\bar{\underline{b}}(\hat{\underline{\Theta}})$ is the same as that of $\bar{\underline{b}}(\hat{\underline{\theta}})$. This does not mean that the estimator $\bar{\underline{b}}(\hat{\underline{\Theta}})$ cannot be more efficient than $\bar{\underline{b}}(\hat{\underline{\theta}})$ in small samples when the disturbances are truly autocorrelated. We are not able to derive the small sample properties of $\bar{\underline{b}}(\hat{\underline{\theta}})$ and $\bar{\underline{b}}(\hat{\underline{\Theta}})$. We will show in Chapter V that the expectation of an estimator of Δ critically depends upon what we assume about the variance-covariance matrix of $\underline{u}_i'$s. If we misspecify the latter there is a high probability that we obtain implausible estimates for the elements of Δ. We may be justified in rejecting a set of assumptions if it leads to implausible estimates. For example, if the estimator (4.3.24) yields negative estimates for variances and the estimator (4.4.9) yields positive estimates for variances, then we may adopt Assumption 4.4.1.

4.5 Problems Associated with the Estimation of RCR Models Using Aggregate Data

So far we have been discussing the inference procedures which can be adopted only when panel data are available. In general, data are available in an aggregate form. Suppose we have data on only macro variables $\bar{\underline{y}}$ and $\bar{X}$ which are obtained by simple aggregation of micro variables, i.e.,

$$\bar{\underline{y}} = \sum_{i=1}^{n} \underline{y}_i \quad \text{and} \quad \bar{X} = \sum_{i=1}^{n} X_i \quad . \tag{4.5.1}$$

If we sum the relationship in (4.2.1) over i we obtain

$$\bar{\underline{y}} = \bar{X}\bar{\underline{\beta}} + \sum_{i=1}^{n} X_i \underline{\delta}_i + \bar{\underline{u}} \tag{4.5.2}$$

where $\bar{\underline{u}} = \sum_{i=1}^{n} \underline{u}_i$.

Let us assume following the usual practice that

$$E\bar{\underline{u}} = 0 \quad \text{and} \quad E\,\bar{\underline{u}}\,\bar{\underline{u}}' = \sigma^2 I_T \quad . \tag{4.5.3}$$

When the specifications (1), (2), (4) and (5) in Assumption 4.2.1 are true, an efficient estimator of $\bar{\underline{\beta}}$ in (4.5.2) is

$$\hat{\bar{\underline{\beta}}}_a = \left[\bar{X}'\Big(\sum_{i=1}^{n} X_i \Delta X_i' + \sigma^2 I\Big)^{-1} \bar{X}\right]^{-1} \bar{X}'\left(\sum_{i=1}^{n} X_i \Delta X_i' + \sigma^2 I\right)^{-1} \bar{\underline{y}} \; . \tag{4.5.4}$$

This estimator can be adopted only if we know X_i $(i=1,2,\ldots,n)$, Δ and σ^2.

If we form the macro estimator

$$\hat{\bar{\underline{\beta}}}_m = (\bar{X}'\bar{X})^{-1}\bar{X}'\bar{\underline{y}} \tag{4.5.5}$$

we can show that

$$E\hat{\bar{\underline{\beta}}}_m = E(\bar{X}'\bar{X})^{-1}\bar{X}'(\bar{X}\bar{\underline{\beta}} + \sum_{i=1}^{n} X_i' \underline{\delta}_i + \bar{\underline{u}}) = \bar{\underline{\beta}} \quad . \tag{4.5.6}$$

Thus, $\bar{\underline{\beta}}_m$ which is the same as (1.2.29), is an unbiased estimator of $\bar{\underline{\beta}}$ though it is not an efficient estimator.

The appropriate formula for calculating the variance-covariance matrix of $\hat{\bar{\underline{\beta}}}_m$ is

$$V(\hat{\bar{\underline{\beta}}}_m) = \sum_{i=1}^{n} P_i \Delta P_i' + \sigma^2 (\bar{X}'\bar{X})^{-1} \tag{4.5.7}$$

where $P_i = (\bar{X}'\bar{X})^{-1}\bar{X}'X_i$, i.e., the matrix of coefficients in the "auxiliary regressions" of X_i upon $\bar{X}$.

It is not difficult to see the distinction between the estimators $\hat{\bar{\underline{\beta}}}_a$ and $\bar{\underline{b}}(\theta)$. The former is obtained by minimizing $\left(\sum_{i=1}^{n} X_i \underline{\delta}_i + \bar{\underline{u}}\right)'\left(\sum_{i=1}^{n} X_i \Delta X_i' + \sigma^2 I\right)^{-1} \cdot \left(\sum_{i=1}^{n} X_i \underline{\delta}_i + \bar{\underline{u}}\right)$ and the latter is obtained by minimizing $[D(X)\underline{\delta}+\underline{u}]'H(\underline{\theta})^{-1}[D(X)\underline{\delta}+\underline{u}]$.

4.6 Forecasting With RCR Models

(a) RCR Model (4.2.1)

We wish to derive in this section an optimal forecasting procedure for a RCR model. Consider the model of the type (4.2.1) for a sample of n individual units. We make Assumption 4.2.1. We want to predict a single drawing of the regressand of the model (4.2.1) given the vector of regressors for the prediction period. The actual drawing will be

$$\underset{(1\times1)}{y^*_{Fi}} = \underset{(1\times K)}{\underline{x}'_{Fi}} \; \underset{(K\times1)}{\underline{\beta}_i} + \underset{(1\times1)}{u_{Fi}} \qquad (1=1,2,\ldots,n) \qquad (4.6.1)$$

where y^*_{Fi} is the scalar value of the regressand for the ith unit, $\underline{x}'_{Fi}$ is the vector of prediction regressors and u_{Fi} is the scalar value of the prediction disturbance. We denote the values of the variables in the forecast period by the subscript F. The drawing on $\underline{\beta}_i$ does not change from one observation to another in the prediction period as in the sample period. Given a sample of n units we will have n drawings on the coefficient vector and these will not change unless we draw a fresh sample of units. As long as we work with a single sample of units the drawings on the coefficient vector remain the same even in the prediction period.

Let Assumption 4.2.1 be true. The second assumption which we shall make is

Assumption 4.6.1: For $i,j=1,2,\ldots,n$, $Eu_{Fi} = 0$, $Eu^2_{Fi} = \sigma_{ii}$, $Eu_{Fi}u_{Fj} = 0$ for all $i \neq j$, $Eu_{Fi}\underline{u}_i = 0$, and $\underline{\beta}_i$ and u_{Fj} are independent for all i and j.

Under these assumptions, given $\underline{x}'_{Fi}$, we can predict a value of y_i, say y_{Fi}, by using the estimated regression relationship,

$$\underline{y}_i = X_i\underline{\bar{b}}(\underline{\theta}) + X_i\underline{\hat{\delta}}_i + \underline{\hat{u}}_i \qquad (i=1,2,\ldots,n) \qquad (4.6.2)$$

where $\underline{\bar{b}}(\underline{\theta})$ is the BLUE of $\underline{\bar{\beta}}$ and $\underline{\hat{\delta}}_i = \underline{b}_i - \underline{\bar{b}}(\underline{\theta})$ is an estimate of the random component in $\underline{\beta}_i = \underline{\bar{\beta}} + \underline{\delta}_i$.

We consider

$$y_{Fi} = \underline{x}'_{Fi}\underline{\bar{b}}(\underline{\theta}) + \underline{x}'_{Fi}[\underline{b}_i - \underline{\bar{b}}(\underline{\theta})] \qquad (i=1,2,\ldots,n) \ . \qquad (4.6.3)$$

The actual value, say y^*_{Fi}, in this prediction period is given in (4.6.1). When we subtract (4.6.1) from (4.6.3) we obtain as the error of forecast

$$\begin{aligned} y_{Fi} - y^*_{Fi} &= \underline{x}'_{Fi}\underline{b}_i - \underline{x}'_{Fi}\underline{\beta}_i - u_{Fi} \\ &= \underline{x}'_{Fi}[\underline{b}_i - \underline{\bar{\beta}}] - \underline{x}'_{Fi}[\underline{\beta}_i - \underline{\bar{\beta}}] - u_{Fi} \qquad (i=1,2,\ldots,n) \ . \qquad (4.6.4) \end{aligned}$$

The expected error is

$$E(y_{Fi} - y^*_{Fi}) = 0 \qquad (i=1,2,\ldots,n) \ , \qquad (4.6.5)$$

because $E\underline{b}_i = \underline{\bar{\beta}}$, $E\underline{\beta}_i = \underline{\bar{\beta}}$, $Eu_{Fi} = 0$, and the $\underline{x}_{Fi}$'s are assumed to be nonstochastic.

We notice that the forecast error in (4.6.4) may be divided into three parts. The first source of error is due to the inaccuracies in estimating the means of regression coefficients as a linear function of $\underline{y}_i$. The second source of error is the deviation of a drawing on $\underline{\beta}_i$ from its mean and the third one is due to the presence of the random term u_{Fi}. Thus the error of forecast is a function of three random variables, $\underline{b}_i$, $\underline{\beta}_i$, and u_{Fi}, which can be reduced to two by expressing $\underline{b}_i - \underline{\beta}_i = [X'_iX_i]^{-1}X'_i\underline{u}_i$ in (4.6.4).

The variance of the forecast error is given by

$$\begin{aligned} E(y_{Fi} - y^*_{Fi})^2 &= E[\underline{x}'_{Fi}(\underline{b}_i - \underline{\bar{\beta}}) - \underline{x}'_{Fi}(\underline{\beta}_i - \underline{\bar{\beta}}) - u_{Fi}]^2 \\ &= \sigma_{ii}\underline{x}'_{Fi}(X'_iX_i)^{-1}x_{Fi} + \sigma_{ii} \qquad (i=1,2,\ldots,n) \ . \qquad (4.6.6) \end{aligned}$$

The predictor (4.6.3) has minimum variance within the class of all linear unbiased predictors because $\underline{b}_i$ is a BLU predictor of $\underline{\beta}_i$.

Instead of predicting a single drawing on the regressand of the model (4.2.1) given the vector of regressors for the prediction period without changing the drawing on the coefficient vector, we can consider predicting a single drawing on the regressand of the model (4.2.1) given the vector of regressors for a unit which is not included in our original cross-section sample. The actual drawing on the regressand for a year will be

$$\underset{(1\times1)}{y^*_{tf}} = \underset{(1\times K)}{\underline{x}'_{tf}} \underset{(K\times1)}{\underline{\beta}_f} + \underset{(1\times1)}{u_{tf}} \qquad (4.6.7)$$

where y^*_{tf} is the scalar value of the regressand for a unit which is not in the original sample, $\underline{x}'_{tf}$ is the vector of regressors for a year and u_{tf} is the scalar value of the disturbance. We denote the values of the variables for a unit which is outside the sample by subscript f. We shall assume

Assumption 4.6.2: $Eu_{tf} = 0$, $Eu^2_{tf} = \sigma_{ff}$, $Eu_{tf}\underline{u}_i = 0$ for all i, $E\underline{\beta}_f = \underline{\bar{\beta}}$, $E(\underline{\beta}_f - \underline{\bar{\beta}})\cdot(\underline{\beta}_f - \underline{\bar{\beta}})' = \Delta$, $\underline{\beta}_f$ is independent of $\underline{u}_i (i=1,2,\ldots,n)$ and $\underline{u}_f$; and $\underline{\beta}_f$ is independent of $\underline{\beta}_i (i=1,2,\ldots,n)$.

Under Assumptions 4.2.1 and 4.6.2, given $\underline{x}'_{tf}$ for a year, we can predict a value of y_{ti}, say y_{tf}, by using the estimated regression relationship (4.6.2). We consider

$$y_{tf} = \underline{x}'_{tf}\underline{\bar{b}}(\underline{\theta}) \quad . \qquad (4.6.8)$$

The actual value, say y^*_{tf}, is given in (4.6.7). When we subtract (4.6.7) from (4.6.8) we obtain as the error of forecast

$$y_{tf} - y^*_{tf} = -\underline{x}'_{tf}[\underline{\beta}_f - \underline{\bar{b}}(\underline{\theta})] - u_{tf}$$

$$= -\underline{x}'_{tf}(\underline{\beta}_f - \underline{\bar{\beta}}) + \underline{x}'_{tf}[\underline{\bar{b}}(\underline{\theta}) - \underline{\bar{\beta}}] - u_{tf} \quad . \qquad (4.6.9)$$

The expectation of the above error is

$$E(y_{tf} - y^*_{tf}) = 0 \quad , \qquad (4.6.10)$$

because $E\underline{\bar{b}}(\theta) = \underline{\bar{\beta}}$, $E\underline{\beta}_f = \underline{\bar{\beta}}$, $Eu_{tf} = 0$, and the $\underline{x}_{tf}$ is assumed to be nonstochastic.

In the case of (4.6.9) the sources of error are -(1) the deviation of a fresh drawing on $\underline{\beta}_i$ for a unit outside the sample from its mean, (2) the inaccuracies in estimating the means of regression coefficients, and (3) the error term u_{tf}.

The variance of forecast error is given by

$$E(y_{tf} - y^*_{tf})^2 = E[\underline{x}'_{tf}(\underline{\beta}_f - \bar{\underline{\beta}}) - \underline{x}'_{tf}(\bar{\underline{b}}(\underline{\theta}) - \bar{\underline{\beta}}) + u_{tf}]^2$$

$$= \underline{x}'_{tf}\Delta\underline{x}_{tf} + \underline{x}'_{tf}C(\underline{\theta})\underline{x}_{tf} + \sigma_{ff} \quad . \qquad (4.6.11)$$

The predictor (4.6.8) has minimum variance within the class of all unbiased predictors which are linear functions of all $\underline{y}_i (i=1,2,...,n)$ because $\bar{\underline{b}}(\underline{\theta})$ is the BLUE of $\bar{\underline{\beta}}$.

(b) Comparison of Forecasting Procedures Under Fixed and Random Coefficient Approaches

In order to compare the predictive abilities of RCR models with those of fixed coefficient models we list below some of the familiar results on forecasting with the latter models.

(1) Consider the model (4.2.1) under the assumption that the $\underline{\beta}_i (i=1,2,...,n)$ are fixed and $\underline{\beta}_1 = \underline{\beta}_2 = ... = \underline{\beta}_n = \underline{\beta}$, say. We also impose here the restriction that $\sigma_{11} = \sigma_{22} = ... = \sigma_{nn} = \sigma^2$, say. Given $\underline{x}_{Fi} (i=1,2,...,n)$ we can obtain a prediction on y_i, say $\hat{y}_{Fi}$, using the estimated relation,

$$\hat{y}_{Fi} = \underline{x}'_{Fi}\underline{b} \qquad (i=1,2,...,n) \qquad (4.6.12)$$

where $\underline{b} = \left(\sum_{i=1}^{n} X'_i X_i\right)^{-1} \sum_{i=1}^{n} X'_i \underline{y}_i$.

The prediction variance of $\hat{y}_{Fi}$ is given by

$$\sigma^2\left[\underline{x}'_{Fi}\left(\sum_{i=1}^{n} X'_i X_i\right)^{-1}\underline{x}_{Fi} + 1\right] \quad . \qquad (4.6.13)$$

(2) Suppose that we consider the following aggregate equation instead of (4.2.1)

$$\sum_{n=1}^{n}\underline{y}_i = \sum_{i=1}^{n} X_i\underline{\beta} + \sum_{i=1}^{n}\underline{u}_i \quad . \qquad (4.6.14)$$

Here we assume that $\underline{\beta}$ is fixed and the same for all individuals. We also assume that

$$E\underline{u}_i = 0 \quad \text{and} \quad E\underline{u}_i\underline{u}'_j = \begin{cases}\sigma^2 I & \text{if } i=j, \\ 0 & \text{if } i\neq j.\end{cases}$$

Given an aggregate vector of prediction regressors, say $\sum_{i=1}^{n} \underline{x}_{Fi}$, we can obtain

$$\sum_{i=1}^{n} \hat{y}_{Fi} = \sum_{i=1}^{n} \underline{x}_{Fi}' \hat{\underline{\beta}} \quad , \tag{4.6.15}$$

where $\hat{\underline{\beta}} = \left[\sum_{i=1}^{n} X_i' \sum_{i=1}^{n} X_i \right]^{-1} \sum_{i=1}^{n} X_i' \sum_{i=1}^{n} \underline{y}_i$.

Its prediction variance is

$$\sigma^2 \left[\sum_{i=1}^{n} \underline{x}_{Fi}' \left(\sum_{i=1}^{n} X_i' \sum_{i=1}^{n} X_i \right)^{-1} \sum_{i=1}^{n} \underline{x}_{Fi} + n \right] \quad . \tag{4.6.16}$$

(3) Let us now consider the model (4.2.1) under the assumption that the $\underline{\beta}_i$'s are fixed and $\underline{\beta}_i = \underline{\beta}$ for all i=1,2,...,n. Let us assume further that

$$E\underline{u}_i = 0 \text{ and } E\underline{u}_i\underline{u}_j' = \begin{cases} \sigma_{ii} I & \text{if } i=j \\ 0 & \text{if } i \neq j \ . \end{cases}$$

Under these assumptions the optimal predictor is

$$\hat{y}_{Fi}^* = \underline{x}_{Fi}' \hat{\underline{\beta}}(\underline{\sigma}) \tag{4.6.17}$$

where

$$\hat{\underline{\beta}}(\underline{\sigma}) = \left[\sum_{j=1}^{n} \frac{X_j'X_j}{\sigma_{jj}} \right]^{-1} \left[\sum_{i=1}^{n} \frac{X_i'\underline{y}_i}{\sigma_{ii}} \right] \ .$$

Its prediction variance is

$$\underline{x}_{Fi}' \left[\sum_{i=1}^{n} \frac{X_i'X_i}{\sigma_{ii}} \right]^{-1} \underline{x}_{Fi} + \sigma_{ii} \quad . \tag{4.6.18}$$

(4) Finally, we treat the $\underline{\beta}_i$'s in (4.2.1) as fixed but different for different individuals. We assume that $E\underline{u}_i = 0$ and

$$E\underline{u}_i\underline{u}_j' = \begin{cases} \sigma_{ii} I & \text{if } i=j, \\ 0 & \text{if } i \neq j. \end{cases}$$

Then the optimal prediction is given by

$$\tilde{y}_{Fi} = \underline{x}_{Fi}' \underline{b}_i \qquad (i=1,2,\ldots,n) \tag{4.6.19}$$

where $\underline{b}_i = (X_i'X_i)^{-1} X_i'\underline{y}_i$.

The prediction variance is given by

$$\sigma_{ii}[\underline{x}'_{Fi}(X'_iX_i)^{-1}x_{Fi} + 1] \qquad (i=1,2,\ldots,n) \quad . \tag{4.6.20}$$

By comparing (4.6.3) with (4.6.19) we find that the optimal forecasting procedure is the same whether the coefficient vector in a microrelation is random across units or fixed and different for different individuals. In both cases we have the same prediction with the same prediction variance. By comparing (4.6.12) with (4.6.17) we can see that the optimal forecasting procedure varies even for a fixed coefficient model depending upon what we assume about the form of the variance-covariance matrix of micro units' disturbance terms. Comparison of (4.6.13) with (4.6.16) indicates that the predictive efficiency of an aggregate relation is different from that of a micro relation. The predictor (4.6.8) is different from the rest and it is relevant only to RCR models.

4.7 Relaxation of Assumptions Underlying RCR Models

In some situations we may find that the specification (4) in Assumption 4.2.1 is very restrictive. We can relax this specification in the following manner. Suppose that G subsamples $S_1,\ldots,S_G$, of sizes $n_1,\ldots,n_G$ units respectively ($\Sigma n_g = n$) are drawn such that each $n_g \geq K + 1$.

Assumption 4.7.1: The $\underline{\beta}_i$ $(i=1,2,\ldots,n)$ are independently distributed with a common mean $\bar{\underline{\beta}}$ and with $E(\underline{\beta}_i-\bar{\underline{\beta}})(\underline{\beta}_i-\bar{\underline{\beta}})' = \Delta_g$ if $i\varepsilon S_g$ $(g=1,2,\ldots,G)$. The specifications (1), (2), (3) and (5) in Assumption 4.2.1 remain as they are.

The above assumption implies that a sample of n cross-section units is drawn from G different populations of $\underline{\beta}_i$. The means of these populations are the same but their variance-covariance matrices are different. A sub-sample of n_g units is drawn from the gth population of $\underline{\beta}_i$. The sizes of these sub-samples are fixed in such a way that no single sub-sample will have a size less than K+1. Using T observations on each of n_g units we can estimate Δ_g by

$$\hat{\Delta}_g = \frac{1}{(n_g-1)}\left[\sum_{i\epsilon S_g} \underline{b}_i\underline{b}_i' - \frac{1}{n_g}\sum_{i\epsilon S_g}\underline{b}_i \sum_{i\epsilon S_g}\underline{b}_i'\right]$$

$$- \frac{1}{n_g}\sum_{i\epsilon S_g} s_{ii}(X_i'X_i)^{-1} \qquad (g=1,2,\ldots,G) \quad . \qquad (4.7.1)$$

A pooled estimator of $\underline{\bar{\beta}}$ is

$$\underline{\bar{b}}(\underline{\hat{\theta}}*) = \left[\sum_{g=1}^{G}\sum_{i\epsilon S_g}\left\{\hat{\Delta}_g + s_{ii}(X_i'X_i)^{-1}\right\}^{-1}\right]^{-1}$$

$$\cdot\left[\sum_{g=1}^{G}\sum_{i\epsilon S_g}\left\{\hat{\Delta}_g + s_{ii}(X_i'X_i)^{-1}\right\}^{-1}\underline{b}_i\right] \qquad (4.7.2)$$

where $\underline{\theta}^*$ is a vector containing all the distinct elements of Δ_g $(g=1,2,\ldots,G)$, and σ_{ii} $(i=1,2,\ldots,n)$.

It is easy to show that

$$\underset{T\to\infty}{\text{plim}}\ |\hat{\Delta}_g - (n_g-1)^{-1}S_\beta(g)| = 0 \qquad (4.7.3)$$

where

$$S_\beta(g) = \sum_{i\epsilon S_g}\underline{\beta}_i\underline{\beta}_i' - \frac{1}{n_g}\sum_{i\epsilon S_g}\underline{\beta}_i\sum_{i\epsilon S_g}\underline{\beta}_i' \quad .$$

Now

$$\underset{T\to\infty}{\text{plim}}\ \underline{\bar{b}}(\underline{\hat{\theta}}^*) = \left[\sum_{g=1}^{G} n_g(n_g-1)S_\beta(g)^{-1}\right]^{-1}\left[\sum_{g=1}^{G}(n_g-1)S_\beta(g)^{-1}\sum_{i\epsilon S_g}\underline{\beta}_i\right] \quad . \qquad (4.7.4)$$

Since $\underset{n_g\to\infty}{\text{plim}}\ \frac{1}{n_g}\sum_{i\epsilon S_g}\underline{\beta}_i = \underline{\bar{\beta}}$, it follows that $\underset{\substack{T\to\infty\\ n_g\to\infty}}{\text{plim}}\ \underline{\bar{b}}(\underline{\hat{\theta}}^*) = \underline{\bar{\beta}}$. Thus $\underline{\bar{b}}(\underline{\hat{\theta}}^*)$ is a consistent estimator of $\underline{\bar{\beta}}$.

If the means of different populations are not the same we cannot pool the data on all the n units and the estimator (4.7.2) contains a specification error. Even the use of aggregated data of the type (1.2.21) leads to biased estimates. To show this let us assume

Assumption 4.7.2: The $\underline{\beta}_i$'s are independently distributed with $E\underline{\beta}_i = \bar{\underline{\beta}}_g$ if $i\epsilon S_g$ $(g=1,2,\ldots,G)$ and $E(\underline{\beta}_i-\bar{\underline{\beta}}_g)(\underline{\beta}_i-\bar{\underline{\beta}}_g)' = \Delta_g$ if $i\epsilon S_g$. The specifications (1), (2), (3) and (5) in Assumption 4.2.1 remain as they are.

Now consider the macro estimator $\hat{\bar{\underline{\beta}}}_m$ in (4.5.5). Following Theil (1954) we can write

$$\hat{\bar{\underline{\beta}}}_m = \frac{1}{n}\sum_{g=1}^{G} n_g\bar{\underline{\beta}}_g + \sum_{g=1}^{G}[(\bar{X}'\bar{X})^{-1}\bar{X}'\sum_{i\epsilon S_g} X_i - \frac{n_g}{n} I]\bar{\underline{\beta}}_g$$

$$+ \sum_{g=1}^{G}(\bar{X}'\bar{X})^{-1}\bar{X}'\sum_{i\epsilon S_g} X_i\underline{\delta}_i + \sum_{g=1}^{G}(\bar{X}'\bar{X})^{-1}\bar{X}'\sum_{i\epsilon S_g}\underline{u}_i \tag{4.7.5}$$

where $\bar{X} = \sum_{g=1}^{G}\sum_{i\epsilon S_g} X_i$ and $\underline{\beta}_i = \bar{\underline{\beta}}_g + \underline{\delta}_i$ if $i\epsilon S_g$.

We call the first term on the r.h.s. of (4.7.5) as the "true value" of macro mean, the second term as the aggregation bias in $\hat{\bar{\underline{\beta}}}_m$ and the sum of last two terms as the implied standard error. Now we can compare (4.7.5) with (1.2.25). Due to Assumption 4.7.2 the implied standard error in (4.7.5) is a function of both micro disturbances and the errors in micro coefficients. In the case of fixed coefficient models the implied standard error is a function of only micro disturbances. The aggregation bias in $\hat{\bar{\underline{\beta}}}_m$ depends upon the covariance between the auxiliary regression coefficient estimators, $(\bar{X}'\bar{X})^{-1}\bar{X}'\Sigma_{i\epsilon S_g}X_i$ and the micro means $\bar{\underline{\beta}}_g$.

4.8 Similarities Between RCR and Bayesian Assumptions

There is a close resemblance between the random coefficient approach as treated in this chapter and a Bayesian approach to fixed coefficient models. For estimation purposes we treated the coefficient vector $\underline{\beta}_i$ as a random variable without specifying the form of its distribution and built an estimator for the mean vector $\bar{\underline{\beta}}$ proceeding conditionally upon a given sample estimate of the variance-covariance matrix $H(\underline{\theta})$, a fact which appears to be the reason why our results are in the main large sample results. A Bayesian, while analyzing a fixed coefficient model with a coefficient vector, say $\underline{\beta}$, would also assume that the

actual value of the coefficient vector in the sampled population was determined by
a random experiment. In such cases, $\underline{\beta}$ is a random variable, having a certain
a priori distribution. He would put a prior pdf, say $p(\underline{\beta}, \sigma^2)$, on the coefficient
vector and the other parameter, say σ^2, of the model. If he had prior knowledge
of the ranges of parameters he would define prior pdf in the appropriate ranges.
He would combine prior pdf with the likelihood function via Bayes' Theorem to
obtain joint posterior density of parameters, i.e.,

$$p(\underline{\beta}, \sigma^2 | \underline{y}) \propto p(\underline{\beta}, \sigma^2) L(\underline{\beta}, \sigma^2 | \underline{y}) \qquad (4.8.1)$$

where $p(\underline{\beta}, \sigma^2 | \underline{y})$ is the joint posterior density of the parameters and $L(\underline{\beta}, \sigma^2 | \underline{y})$ is
the likelihood function. He would integrate out the nuisance parameter σ^2 over the
appropriate range and obtain a posterior density for the coefficient vector. This
posterior density will be useful in making inferences on $\underline{\beta}$, cf. Tiao and Zellner
(1964).

If a Bayesian were to analyze a RCR model he would put prior pdf on the
mean vector $\bar{\underline{\beta}}$ and the vector $\underline{\theta}$ of variances and covariances. Using it in conjunction with the likelihood function he would obtain the posterior density function of
$(\bar{\underline{\beta}}, \underline{\theta})$ as

$$p(\bar{\underline{\beta}}, \underline{\theta} | y) \propto p(\bar{\underline{\beta}}, \underline{\theta}) L(\bar{\underline{\beta}}, \underline{\theta} | \underline{y}) \qquad . \qquad (4.8.2)$$

At this point if his interest centers at the mean vector $\bar{\underline{\beta}}$, he would integrate out
$\underline{\theta}$ over the appropriate range to get the marginal posterior density of $\underline{\beta}$, and thus
avoid the necessity for proceeding conditionally upon a given sample estimate of $\underline{\theta}$.
As mentioned in subsection 4.3(c), the problem of making inferences about the
variances of coefficients gives rise to considerable difficulty in the sampling
theory framework. No such difficulty exists when one analyzes the problem from a
Bayesian point of view because the range of variance can be easily constrained to
be positive in the Bayesian analysis.

4.9 Empirical CES Production Function Free of Management Bias

In subsection 4.2(b) we specified a CES production function with management as one of the inputs. Now we briefly indicate a procedure of estimating the parameters of this function under Assumption 4.2.1. We assume that γ, δ and α are fixed parameters and they are the same for all firms. The parameter ρ_i is fixed but different for different firms. The managerial services varies randomly across firms. This introduces randomness in the coefficients of (4.2.1b). We assume that the managerial services are independent of the disturbance term so that the specification (5) in Assumption 4.2.1 is valid. Adopting Assumption 4.3.1 we can apply the ML method described in Zarembka (1968) to obtain a prediction on the coefficient vector of each firm and the estimates of ρ_i and σ_{ii}. We can substitute these estimates of σ_{ii} and predictions in $\hat{\Delta}$ and $\underline{\bar{b}}(\hat{\underline{\theta}})$, and obtain the estimates of the means of coefficients $\beta_{ki}(k=0,1,2)$. In this approach we cannot identify γ and α separately. We can obtain an estimate of δ.

The management variable in (4.2.1a) is assumed to be invariant over time and this assumption is likely to be true only in the short run. It may appear silly to apply an asymptotic procedure to estimate the equation (4.2.1b). Unfortunately, the analytical study of the small sample properties of $\underline{\bar{b}}(\hat{\underline{\theta}})$ is difficult. Monte Carlo study of this problem may throw some light. Often asymptotic properties provide good insights into the small sample properties. As we indicated in (4.3.16) $\underline{\bar{b}}(\hat{\underline{\theta}})$ is likely to be more efficient than the OLS estimator of $\underline{\bar{\beta}}$ even in small samples if Δ is appreciably different from a null matrix. If Δ and σ_{ii} $(i=1,2,\ldots,n)$ are precisely estimated, $\underline{\bar{b}}(\hat{\underline{\theta}})$ is likely to be more efficient than any other feasible estimator of $\underline{\bar{\beta}}$.

4.10 Analysis of Mixed Models

If the model (4.2.1) contains lagged values of the dependent variable as explanatory variables we cannot make use of the procedure developed in Section 4.3 unless the regressors are independent of their corresponding coefficients and the disturbance terms. The condition that a regressor is independent of its own coefficient appears to be very strong and may be false in many occasions. To avoid this condition in the estimation of a dynamic relation of the type (1.2.1) with random coefficients, we may, if appropriate, assume that the coefficients of lagged values of the dependent variable are nonrandom and only the coefficients of exogenous variables are random. In the case of (1.2.1) this means that $\gamma_{1i}, \gamma_{2i},\ldots,\gamma_{Gi}$ are nonrandom and the coefficients $\beta_{0i}, \delta_{1i}, \delta_{2i}, \ldots, \delta_{K-1,i}$ are random across units. We can either assume that the γ_{gi}'s are fixed and different for different units or assume that $\gamma_{gi} = \gamma_g$ for all i. Under these assumptions we will have a mixed model containing both fixed and random coefficients.

The regression model with both fixed and random coefficients can be represented as

$$\underline{y}_i = X_{1i}\underline{\beta}_{1i} + X_{2i}\underline{\beta}_{2i} + \underline{u}_i \qquad (i=1,2,\ldots,n) \quad , \qquad (4.10.1)$$

where $\underline{y}_i$ is a TX1 vector of observations on the regress-and, X_{1i} is a TXK_1 matrix of observations on K_1 independent variables, $\underline{\beta}_{1i}$ is a K_1X1 vector of nonrandom coefficients, X_{2i} is a TXK_2 matrix of observations on K_2 independent variables, $\underline{\beta}_{2i}$ is a K_2X1 vector of random coefficients, and $\underline{u}_i$ is a TX1 vector of disturbances. We shall assume

Assumption 4.10.1:

(1) The sample size T is greater than $K = K_1 + K_2$.

(2) The X_{1i} and X_{2i} are nonstochastic. The rank of X_{1i} is K_1 and the rank of X_{2i} is K_2.

(3) $\underline{\beta}_{2i}$ (i=1,2,...,n) are independently and identically distributed with mean $\bar{\underline{\beta}}_2$ and the variance-covariance matrix Δ_2.

(4) $\underline{u}_i$ (i=1,2,...,n) are independently distributed with $E\underline{u}_i = 0$ and $E\underline{u}_i\underline{u}_i' = \sigma_{ii}I_T$.

(5) $\underline{\beta}_{2i}$ and $\underline{u}_j$ are independent of each other for every i and j = 1,2,...,n.

(6) $\underline{\beta}_{1i}$ (i=1,2,...,n) are the fixed coefficient vectors.

Now the problem is how to estimate the parameters, $\underline{\bar{\beta}}_2$, Δ_2, $\underline{\beta}_{1i}$, σ_{ii} (i=1,2,...,n) when Assumption 4.10.1 is true and T time series observations on each variable for each of n units are available.

We can write (4.10.1) as

$$\begin{bmatrix} \underline{y}_1 \\ \underline{y}_2 \\ \cdot \\ \cdot \\ \cdot \\ \underline{y}_n \end{bmatrix} = \begin{bmatrix} X_{11} & 0 & \cdots & 0 \\ 0 & X_{12} & \cdots & 0 \\ \cdot & \cdot & & \cdot \\ \cdot & \cdot & & \cdot \\ \cdot & \cdot & & \cdot \\ 0 & 0 & \cdots & X_{1n} \end{bmatrix} \begin{bmatrix} \underline{\beta}_{11} \\ \underline{\beta}_{12} \\ \cdot \\ \cdot \\ \cdot \\ \underline{\beta}_{1n} \end{bmatrix} + \begin{bmatrix} X_{21} \\ X_{22} \\ \cdot \\ \cdot \\ \cdot \\ X_{2n} \end{bmatrix} \underline{\bar{\beta}}_2 + \begin{bmatrix} X_{21} & 0 & \cdots & 0 \\ 0 & X_{22} & \cdots & 0 \\ \cdot & \cdot & & \cdot \\ \cdot & \cdot & & \cdot \\ \cdot & \cdot & & \cdot \\ 0 & 0 & \cdots & X_{2n} \end{bmatrix} \begin{bmatrix} \underline{\delta}_{21} \\ \delta_{22} \\ \cdot \\ \cdot \\ \cdot \\ \underline{\delta}_{2n} \end{bmatrix} + \begin{bmatrix} \underline{u}_1 \\ \underline{u}_2 \\ \cdot \\ \cdot \\ \cdot \\ \underline{u}_n \end{bmatrix} \tag{4.10.2}$$

or

$$\underline{y} = D(X_1)\underline{\beta}_1 + X_2\underline{\bar{\beta}}_2 + D(X_2)\underline{\delta}_2 + \underline{u} \quad , \tag{4.10.2a}$$

where

$\underline{y} \equiv [\underline{y}_1', \underline{y}_2', \ldots, \underline{y}_n']'$, $D(X_1) \equiv \text{diag}[X_{11}, X_{12}, \ldots, X_{1n}]$,

$\underline{\beta}_1 \equiv [\underline{\beta}_{11}', \underline{\beta}_{12}', \ldots, \underline{\beta}_{1n}']'$, $X_2 \equiv [X_{21}', X_{22}', \ldots, X_{2n}']'$,

$D(X_2) \equiv \text{diag}[X_{21}, X_{22}, \ldots, X_{2n}]$, $\underline{\delta}_2 \equiv [\underline{\delta}_{21}', \underline{\delta}_{22}', \ldots, \underline{\delta}_{2n}']'$, and

$\underline{u} \equiv [\underline{u}_1', \underline{u}_2', \ldots, \underline{u}_n']'$.

The variance-covariance matrix of $D(X_2)\underline{\delta} + \underline{u}$ is given by

$$H_2(\underline{\theta}_2) = \begin{bmatrix} X_{21}\Delta_2X_{21}' + \sigma_{11}I_T & 0 & \cdots & 0 \\ 0 & X_{22}\Delta_2X_{22}' + \sigma_{22}I_T & \cdots & 0 \\ \cdot & \cdot & & \cdot \\ \cdot & \cdot & & \cdot \\ \cdot & \cdot & & \cdot \\ 0 & 0 & \cdots & X_{2n}\Delta_2X_{2n}' + \sigma_{nn}I_T \end{bmatrix} \qquad (4.10.3)$$

where $\underline{\theta}_2$ is a $\frac{1}{2}[K_2(K_2+1) + 2n] \times 1$ vector containing all the distinct elements of Δ_2 and $\sigma_{11},\ldots,\sigma_{nn}$. We assume that $H_2(\underline{\theta}_2)$ is nonsingular for all $\underline{\theta}_2 \epsilon A_2$ which is an interval in the $\frac{1}{2}[K_2(K_2+1) + 2n]$ - dimensional space containing the true value of $\underline{\theta}_2$ as an interior point.

Applying Aitken's generalized least squares to (7.4.2a) we obtain the BLUE of $\underline{\beta}_1$ and $\bar{\underline{\beta}}_2$ as

$$\begin{bmatrix} \underline{b}_1(\underline{\theta}_2) \\ \bar{\underline{b}}_2(\underline{\theta}_2) \end{bmatrix} = \begin{bmatrix} D(X_1)'H_2(\underline{\theta}_2)^{-1}D(X_1) & D(X_1)'H_2(\underline{\theta}_2)^{-1}X_2 \\ X_2'H_2(\underline{\theta}_2)^{-1}D(X_1) & X_2'H_2(\underline{\theta}_2)^{-1}X_2 \end{bmatrix}^{-1} \begin{bmatrix} D(X_1)'H_2(\underline{\theta}_2)^{-1}\underline{y} \\ X_2'H_2(\underline{\theta}_2)^{-1}\underline{y} \end{bmatrix}. \qquad (4.10.4)$$

The estimators in (4.10.4) are unbiased and their variance-covariance matrix is

$$\begin{bmatrix} D(X_1)'H_2(\underline{\theta}_2)^{-1}D(X_1) & D(X_1)'H_2(\underline{\theta}_2)^{-1}X_2 \\ X_2'H_2(\underline{\theta}_2)^{-1}D(X_1) & X_2'H_2(\underline{\theta}_2)^{-1}X_2 \end{bmatrix}^{-1} . \qquad (4.10.5)$$

If we do not know the elements of $\underline{\theta}_2$, we can replace them by their unbiased estimators. We can estimate σ_{ii} by s_{ii} in (4.3.21) and Δ_2 by

$$\hat{\Delta}_2 = \frac{S_{b_2}}{n-1} - \frac{1}{n}\sum_{i=1}^{n} s_{ii}(X_{2i}'M_{1i}X_{2i})^{-1} , \qquad (4.10.6)$$

where

$$S_{b_2} = \sum_{i=1}^{n} \underline{b}_{2i}\underline{b}_{2i}' - \frac{1}{n}\sum_{i=1}^{n} \underline{b}_{2i} \sum_{i=1}^{n} \underline{b}_{2i}' ,$$

$$\underline{b}_{2i} = (X_{2i}'M_{1i}X_{2i})^{-1}X_{2i}'M_{1i}\underline{y}_i ,$$

$$M_{1i} = I_T - X_{1i}(X_{1i}'X_{1i})^{-1}X_{1i}' .$$

$\hat{\Delta}_2$ is an unbiased estimator of Δ_2.

Replacing Δ_2 and $\sigma_{ii}(i=1,2,\ldots,n)$ in (4.10.4) by $\hat{\Delta}_2$ and $s_{ii}(i=1,2,\ldots,n)$, respectively, we obtain feasible estimators $\underline{b}_1(\hat{\underline{\theta}}_2)$ and $\underline{\bar{b}}_2(\hat{\underline{\theta}}_2)$ say, for $\underline{\beta}_1$ and $\underline{\bar{\beta}}_2$, respectively. Following the standard procedures discussed in subsection 4.3(g) we can show that $\underline{b}_1(\hat{\underline{\theta}}_2)$ and $\underline{\bar{b}}_2(\hat{\underline{\theta}}_2)$ are consistent and asymptotically efficient.

Extension of the above procedure to cases where $\underline{\beta}_{11} = \underline{\beta}_{12} = \ldots = \underline{\beta}_{1n} = \underline{\bar{\beta}}_1$, say, is straightforward.

A special case of the model (4.10.1) is when $X_{1i} \equiv \underline{\iota}_T$. If all the independent variables in (1.2.4) are nonstochastic and the μ_i's are fixed and different for different units, then the model (1.2.4) is the same as the model (4.10.1) provided $X_{1i} \equiv \underline{\iota}_T$, $\Delta_2 = 0$ and $\sigma_{11} = \sigma_{22} = \ldots = \sigma_{nn}$.

Notice that the test of hypothesis Δ=0 in subsection 4.3(1) is useful only in making a choice between the assumption that the coefficient vectors $\underline{\beta}_i$ are random in the sense of (4.2.1) and the assumption that the coefficient vectors $\underline{\beta}_i$ are fixed and equal for all units. It is entirely possible that for a given situation neither of these two assumptions is correct and the coefficient vectors of different units come from different distributions, or distributions with different means, which would still provide random coefficients. Suppose that for i=1,2,...,n; $E\underline{\beta}_i = \underline{\bar{\beta}}_i$ and $E(\underline{\beta}_i - \underline{\bar{\beta}}_i)(\underline{\beta}_j - \underline{\bar{\beta}}_j)' = \bar{\Delta}$ if $i=j$; 0 if $i \neq j$. In this case the restriction that $\bar{\Delta}=0$ implies that the coefficient vectors $\underline{\beta}_i$ are nonrandom but different for different units. There is a systematic variation in $\underline{\beta}_i$ across units. Data on each unit should be used to estimate its own fixed coefficient vector. If the restriction $\bar{\Delta}=0$ does not hold, we cannot estimate $\bar{\Delta}$ with data of the form given in (4.2.1). We can only estimate the means of coefficient vectors. After estimating n different means of coefficient vectors and n different variances of disturbances, we do not have anymore degrees of freedom to estimate $\bar{\Delta}$. Furthermore, we cannot pool the data on all units to estimate either the means or the variances. As we have already shown in Section 4.7 a single macro relation cannot be derived by aggregation from a set of micro relations of the form (4.2.1) if we define the macro variables as in (1.2.21) and the "true" values of macro coefficients as in (1.2.25). A sufficient condition for pooling the data on all units to estimate a coefficient vector is that $E\underline{\beta}_i = \bar{\underline{\beta}}$ for all i. We have also shown in subsection 4.6(a) that an optimal procedure of predicting a drawing on the regressand for the prediction period with the model (4.2.1) under Assumption 4.2.1 is the same as that with the model (4.2.1) under the assumption that the $\underline{\beta}_i$'s are fixed and different for different units. On the basis of the data given in (4.2.1) we cannot discriminate between the assumption that coefficient vectors are random across units with the same mean and the assumption that coefficient vectors are fixed but different for different units. However, there are other practical procedures for detecting the specification errors in Assumption 4.2.1. If we obtain implausible estimates for the parameters when we estimate (4.2.1) under Assumption 4.2.1, we may doubt the validity of some of the specifications in Assumption 4.2.1. The discussion in subsection 3.12(b) clearly indicates that the inaccuracies in the specifying assumptions could lead to implausible estimates.

We have noted in subsection 4.3(c) that the estimator $\hat{\Delta}$ can yield negative estimates for the variances of coefficients with positive probability. It is possible to impose the restriction that Δ is positive semi-definite by using methods described by Nelder (1968) for simpler cases. Instead, we may prefer to take the negative estimates of variances as indicators of errors in our specification of either the model or the underlying assumptions. As can be noticed from (4.3.24), an unbiased estimator of Δ is in the form of a difference between two matrices. The second matrix on the r.h.s. of (4.3.24) depends upon what we assume about the form of the variance-covariance matrix of $\underline{u}_i$'s. If we assume homoskedasticity of disturbances imposing the restriction, $\sigma_{11} = \sigma_{22} = \dots = \sigma_{nn}$, an unbiased estimator of Δ is given by

$$\hat{\Delta}^* = \frac{S_b}{n-1} - \frac{\hat{\sigma}^2}{n} \sum_{i=1}^{n} (X_i'X_i)^{-1} \tag{4.10.7}$$

where $E\underline{u}_i\underline{u}_j' = \begin{cases} \sigma^2 I & \text{if } i=j \\ 0 & \text{if } i \neq j \end{cases}$ and $\hat{\sigma}^2 = \left(\sum_{i=1}^{n} \underline{y}_i' M_i \underline{y}_i\right)/n(T-K)$. It is possible that for a given sample the diagonal elements of $\frac{\hat{\sigma}^2}{n} \sum_{i=1}^{n} (X_i'X_i)^{-1}$ will be larger than the corresponding diagonal elements of $\frac{1}{n} \sum_{i=1}^{n} s_{ii}(X_i'X_i)^{-1}$ and the estimator (4.10.7) yields negative estimates for the variances of coefficients whereas the estimator (4.3.24) yields positive estimates for the same variances.

Secondly, a possible cause of negative estimated variance is the following: Suppose that we make the specification (3) in Assumption 4.2.1 when in fact Assumption 4.4.1 is operative. Due to this inaccuracy we use the formula (4.3.24) for estimating Δ instead of the appropriate formula. The appropriate formula for estimating Δ under Assumption 4.4.1 is

$$\hat{\Delta}_\rho = \frac{S_b}{n-1} - \frac{1}{n} \sum_{i=1}^{n} \sigma_{ii}(X_i'X_i)^{-1}X_i'\Omega_{ii}X_i(X_i'X_i)^{-1} \quad . \tag{4.10.8}$$

It was shown in Goldberger (1964, pp. 238-42) that the diagonal elements of $\sigma_{ii}(X_i'X_i)^{-1}$ are larger than the corresponding diagonal elements of $\sigma_{ii}(X_i'X_i)^{-1} \cdot X_i'\Omega_{ii}X_i(X_i'X_i)^{-1}$ if the elements of $\underline{u}_i$ are negatively autocorrelated and the regressors, x_{kit}, are positively autocorrelated. In this case we will be subtracting more than what we should from the diagonal elements of $(n-1)^{-1}S_b$ and we might get negative estimates for the variances. Since positive autocorrelation is the typical situation in economic time series it is highly improbable that for any given sample the estimator (4.3.24) yields negative estimates due to the above inaccuracy alone.

Thirdly, it may happen that for a given set of data, the specification (3) of Assumption 4.2.1 is incorrect. The following assumption may be the right one.[19]

Assumption 4.10.2:

$E\underline{u}_i = 0$ for all i and $E\underline{u}_i\underline{u}_j' = \sigma_{ij}I_T$ for all i&j. In this case, the correct formula for estimating Δ is

[19] The statistical dependence between $\underline{u}_i$ and $\underline{u}_j$ for $i \neq j$ may be due to nonreplacement sampling from an existing finite population, cf. McHugh and Mielke (1968). For the same reason $\underline{\beta}_i$ and $\underline{\beta}_j$ for $i \neq j$ may be correlated. To avoid some complications we maintain the specification (4) in Assumption 4.2.1. The "fact" of the matter is that economic data are not usually obtained by random sampling procedures. Whichever assumption one adopts, one is simplifying.

$$\frac{S_b}{n-1} - \frac{1}{n}\sum_{i=1}^{n} s_{ii}(X_i'X_i)^{-1} + \frac{1}{n(n-1)}\sum_{i\neq j} s_{ij}(X_i'X_i)^{-1}X_i'X_j(X_j'X_j)^{-1} \qquad (4.10.9)$$

where $s_{ij} = \underline{y}_i'M_iM_j\underline{y}_j/\text{tr } M_iM_j$ and M_i is as shown in (4.3.5). We can easily show that (4.10.9) is an unbiased estimator of Δ when the specifications (2) and (4) in Assumption 4.2.1 and Assumption 4.10.2 are true. If majority of covariance estimates $s_{ij}(i\neq j)$ are positive and the diagonal elements of $X_i'X_j$ for many pairs of individual units are positive, then the diagonal elements of the third matrix in (4.10.9) will be positive. Consequently, there could be situations in which the estimator (4.10.9) yields positive estimates for the variances and the estimator (4.3.24) yields negative estimates for the same variances.

If an unbiased estimator of Δ under a reasonable assumption about the form of the variance-covariance matrix of $\underline{u}_i$'s yields negative estimates for the variances of some coefficients, it may be desirable to assume that the regression coefficients corresponding to the negative estimated variances are not variable as the $\underline{\beta}_{1i}$'s in (4.10.1). These variances and the corresponding covariances would then be zero and can be removed from Δ.

Now we develop asymptotic procedures for the test of hypotheses: (1) $\Delta_2 = 0$ given that $E\underline{\beta}_{2i} = \bar{\underline{\beta}}_2$, and (2) $\Delta_2 = 0$ and $\underline{\beta}_{1i} = \bar{\underline{\beta}}_1$ given that $E\underline{\beta}_{2i} = \bar{\underline{\beta}}_2$.

Under Assumption 4.10.1 and the assumption that $D(X_2)\underline{\delta}_2 + \underline{u}$ is normally distributed the likelihood function of parameters in (4.10.2a) is

$$L(\underline{\beta}_1,\bar{\underline{\beta}}_2,\underline{\theta}_2|\underline{y},X) = (2\pi)^{-nT/2}\prod_{i=1}^{n}\left\{\sigma_{ii}^{-\frac{1}{2}(T-K_2)}|X_{2i}'X_{2i}|^{-\frac{1}{2}}|\Delta_2 + \sigma_{ii}(X_{2i}'X_{2i})^{-1}|^{-\frac{1}{2}}\right\}$$

$$\cdot \exp\left\{-\frac{1}{2}\sum_{i=1}^{n}\left[(\underline{y}_i - X_{1i}\underline{\beta}_{1i})'\frac{M_{2i}}{\sigma_{ii}}(\underline{y}_i - X_{1i}\underline{\beta}_{1i})\right.\right.$$

$$+ (\hat{\underline{\beta}}_{2i} - \hat{X}_{1i}\underline{\beta}_{1i} - \bar{\underline{\beta}}_2)'[\Delta_2 + \sigma_{ii}(X_{2i}'X_{2i})^{-1}]^{-1}$$

$$\left.\left.\cdot(\hat{\underline{\beta}}_{2i} - \hat{X}_{1i}\underline{\beta}_{1i} - \bar{\underline{\beta}}_2)\right]\right\} \qquad (4.10.10)$$

where $\hat{\underline{\beta}}_{2i} = (X_{2i}'X_{2i})^{-1}X_{2i}'\underline{y}_i$, $\hat{X}_{1i} = (X_{2i}'X_{2i})^{-1}X_{2i}'X_{1i}$, $M_{2i} = I_T - X_{2i}(X_{2i}'X_{2i})^{-1}X_{2i}'$ and use is made of the results $|X_{2i}\Delta_2X_{2i}' + \sigma_{ii}I_T| = \sigma_{ii}^{(T-K_2)}|X_{2i}'X_{2i}||\Delta_2 + \sigma_{ii} \cdot (X_{2i}'X_{2i})^{-1}|$ and $(X_{2i}\Delta_2X_{2i}' + \sigma_{ii}I_T)^{-1} = \sigma_{ii}^{-1}M_{2i} - X_{2i}(X_{2i}'X_{2i})^{-1}[\Delta_2 + \sigma_{ii}(X_{2i}'X_{2i})^{-1}]^{-1} \cdot (X_{2i}'X_{2i})^{-1}X_{2i}'$.

Differentiating $\log L(\underline{\beta}_1, \overline{\underline{\beta}}_2, \underline{\theta}_2|\underline{y}, X)$ with respect to $\underline{\beta}_{1i}$ and equating it to zero we have

$$\frac{\partial \log L}{\partial \underline{\beta}_{1i}} = \frac{1}{\sigma_{ii}}\left[X_{1i}'M_{2i}\underline{y}_i - X_{1i}'M_{2i}X_{1i}\underline{\beta}_{1i}\right] + \hat{X}_{1i}'\left[\Delta_2 + \sigma_{ii}(X_{2i}'X_{2i})^{-1}\right]^{-1} \cdot (\hat{\underline{\beta}}_{2i} - \hat{X}_{1i}\underline{\beta}_{1i} - \overline{\underline{\beta}}_2) = 0 \qquad (i=1,2,\ldots,n) \quad . \tag{4.10.11}$$

Solving (4.10.11) for $\underline{\beta}_{1i}$ in terms of $\overline{\underline{\beta}}_2$ and $\underline{\theta}_2$ we have

$$\underline{b}_{1i}(\overline{\underline{\beta}}_2, \underline{\theta}_2) = \left\{\frac{X_{1i}'M_{2i}X_{1i}}{\sigma_{ii}} + \hat{X}_{1i}'[\Delta_2 + \sigma_{ii}(X_{2i}'X_{2i})^{-1}]^{-1}\hat{X}_{1i}\right\}^{-1} \cdot \left\{\frac{X_{1i}'M_{2i}\underline{y}_i}{\sigma_{ii}} + \hat{X}_{1i}'[\Delta_2 + \sigma_{ii}(X_{2i}'X_{2i})^{-1}]^{-1} \cdot (\hat{\underline{\beta}}_{2i} - \overline{\underline{\beta}}_2)\right\} \qquad (i=1,2,\ldots,n) \quad . \tag{4.10.12}$$

Differentiating $\log L(\underline{\beta}_1, \overline{\underline{\beta}}_2, \underline{\theta}_2|\underline{y}, X)$ with respect to $\overline{\underline{\beta}}_2$ and equating it to zero we have

$$\frac{\partial \log L}{\partial \overline{\underline{\beta}}_2} = \sum_{i=1}^{n} \left\{\Delta_2 + \sigma_{ii}(X_{2i}'X_{2i})^{-1}\right\}^{-1} (\hat{\underline{\beta}}_{2i} - \hat{X}_{1i}\underline{\beta}_{1i} - \overline{\underline{\beta}}_2) = 0 \quad . \tag{4.10.13}$$

Solving (4.10.13) for $\overline{\underline{\beta}}_2$ in terms of $\underline{\beta}_{1i}$ and $\underline{\theta}_2$ we have

$$\overline{\underline{b}}_2(\underline{\beta}_1, \underline{\theta}_2) = \left[\sum_{j=1}^{n} \left\{\Delta_2 + \sigma_{jj}(X_{2j}'X_{2j})^{-1}\right\}^{-1}\right]^{-1} \cdot \left[\sum_{i=1}^{n} \left\{\Delta_2 + \sigma_{ii}(X_{2i}'X_{2i})^{-1}\right\}^{-1} (\hat{\underline{\beta}}_{2i} - \hat{X}_{1i}\underline{\beta}_{1i})\right] . \tag{4.10.14}$$

If we solve the equations (4.10.11) and (4.10.13) simultaneously for $\underline{\beta}_1$ and $\overline{\underline{\beta}}_2$ in terms of $\underline{\theta}$ we obtain (4.10.4) as an estimate of $(\underline{\beta}_1', \overline{\underline{\beta}}_2')'$.

Differentiating $\log L(\underline{\beta}_1,\underline{\bar{\beta}}_2,\underline{\theta}_2|\underline{y},X)$ with respect to σ_{ii} we have

$$\frac{\partial \log L}{\partial \sigma_{ii}} = -\frac{1}{2}\frac{(T-K_2)}{\sigma_{ii}} - \frac{1}{2}\operatorname{tr}[\Delta_2 + \sigma_{ii}(X_{2i}'X_{2i})^{-1}]^{-1}(X_{2i}'X_{2i})^{-1}$$

$$+ \frac{1}{2}\frac{(\underline{y}_i - X_{1i}\underline{\beta}_{1i})'M_{2i}(\underline{y}_i - X_{1i}\underline{\beta}_{1i})}{\sigma_{ii}^2}$$

$$+ \frac{1}{2}\operatorname{tr}[\Delta_2 + \sigma_{ii}(X_{2i}'X_{2i})^{-1}]^{-1}(\hat{\underline{\beta}}_{2i} - \hat{X}_{1i}\underline{\beta}_{1i} - \underline{\bar{\beta}}_2)(\hat{\underline{\beta}}_{2i} - \hat{X}_{1i}\underline{\beta}_{1i} - \underline{\bar{\beta}}_2)'$$

$$\cdot [\Delta_2 + \sigma_{ii}(X_{2i}'X_{2i})^{-1}]^{-1}(X_{2i}'X_{2i})^{-1} \quad (i=1,2,\ldots,n). \qquad (4.10.15)$$

Similarly, differentiating $\log L(\underline{\beta}_1,\underline{\bar{\beta}}_2,\underline{\theta}_2\ \underline{y},X)$ with respect to Δ_2 we have

$$\frac{\partial \log L}{\partial \Delta_2} = -\frac{1}{2}\sum_{i=1}^{n}[\Delta_2 + \sigma_{ii}(X_{2i}'X_{2i})^{-1}]^{-1} + \frac{1}{2}\sum_{i=1}^{n}[\Delta_2 + \sigma_{ii}(X_{2i}'X_{2i})^{-1}]^{-1}$$

$$\cdot (\hat{\underline{\beta}}_{2i}-\hat{X}_{1i}\underline{\beta}_{1i}-\underline{\bar{\beta}}_2)(\hat{\underline{\beta}}_{2i}-\hat{X}_{1i}\underline{\beta}_{1i}-\underline{\bar{\beta}}_2)'[\Delta_2 + \sigma_{ii}(X_{2i}'X_{2i})^{-1}]^{-1} \quad . \qquad (4.10.16)$$

Equating (4.10.15) and (4.10.16) to zero and replacing $\underline{\beta}_1$ and $\underline{\bar{\beta}}_2$ by $\underline{b}_1(\underline{\theta}_2)$ and $\underline{\bar{b}}_2(\underline{\theta}_2)$ in (4.10.4) respectively, we obtain $n+K_2(K_2+1)/2$ equations in terms of $\underline{\theta}_2$ and observations. We can solve these equations by the steepest ascent method developed by Hartley and Rao (1967). Let $\underline{b}_1(\tilde{\underline{\theta}}_2)$, $\underline{\bar{b}}_2(\tilde{\underline{\theta}}_2)$, $\tilde{\sigma}_{ii}$ $(i=1,2,\ldots,n)$ and $\tilde{\Delta}_2$ be the solutions of the equations (4.10.11), (4.10.13), (4.10.15) and (4.10.16) respectively. Consequently

$$\sup_{\underline{\beta}_1,\underline{\bar{\beta}}_2,\underline{\theta}_2} L(\underline{\beta}_1,\underline{\bar{\beta}}_2,\underline{\theta}_2|\underline{y},X) = L(\underline{b}_1(\tilde{\underline{\theta}}_2),\ \underline{\bar{b}}_2(\tilde{\underline{\theta}}_2),\tilde{\underline{\theta}}_2|\underline{y},X) \quad . \qquad (4.10.17)$$

When $\Delta_2 = 0$,

$$L(\underline{\beta}_1,\underline{\bar{\beta}}_2,\underline{\theta}_2|\underline{y},X) = (2\pi)^{-nT/2}\prod_{i=1}^{n}\sigma_{ii}^{-T/2}$$

$$\cdot \exp\left\{-\frac{1}{2}\left[\sum_{i=1}^{n}\frac{(\underline{y}_i-X_{1i}\underline{\beta}_{1i}-X_{2i}\underline{\bar{\beta}}_2)'(\underline{y}_i-X_{1i}\underline{\beta}_{1i}-X_{2i}\underline{\bar{\beta}}_2)}{\sigma_{ii}}\right]\right\} . \qquad (4.10.18)$$

The ML estimates of $\underline{\beta}_1$ and $\bar{\underline{\beta}}_2$ in terms of $\underline{\sigma}$ when $\Delta_2=0$ are given by

$$\begin{bmatrix} \underline{b}_1(\underline{\sigma}) \\ \\ \bar{\underline{b}}_2(\underline{\sigma}) \end{bmatrix} = \begin{bmatrix} D'(X_1)\Sigma_d^{-1}D(X_1) & D'(X_1)\Sigma_d^{-1}X_2 \\ \\ X_2'\Sigma_d^{-1}D(X_1) & X_2'\Sigma_d^{-1}X_2 \end{bmatrix}^{-1} \begin{bmatrix} D'(X_1)\Sigma_d^{-1}\underline{y} \\ \\ X_2'\Sigma_d^{-1}\underline{y} \end{bmatrix} \tag{4.10.19}$$

where $\Sigma_d = \text{diag}[\sigma_{11}I_T,\sigma_{22}I_T,\ldots,\sigma_{nn}I_T]$. The ML estimates of σ_{ii} when $\Delta_2 = 0$, are given by

$$\hat{\sigma}^*_{ii} = \frac{1}{T}\,[\underline{y}_i - X_{1i}\underline{b}_{1i}(\underline{\sigma}) - X_{2i}\bar{\underline{b}}_2(\underline{\sigma})]'[\underline{y}_i - X_{1i}\underline{b}_{1i}(\underline{\sigma}) - X_{2i}\bar{\underline{b}}_2(\sigma)] \tag{4.10.20}$$

$$(i=1,2,\ldots,n)$$

where $\underline{b}_1(\underline{\sigma}) = [\underline{b}_{11}'(\underline{\sigma}),\ \underline{b}_{12}'(\underline{\sigma}),\ \ldots,\ \underline{b}_{1n}'(\underline{\sigma})]'$.

Starting with consistent estimates of σ_{ii} we can evaluate (4.10.19) and (4.10.20) iteratively till we obtain stable estimates for $\underline{\beta}_1$, $\bar{\underline{\beta}}_2$ and $\underline{\sigma}$. Let $[\underline{b}_1'(\hat{\underline{\sigma}}),\ \bar{\underline{b}}_2'(\hat{\underline{\sigma}})]'$ and $\hat{\sigma}^*_{ii}$ $(i=1,2,\ldots,n)$ be the ML estimates of $\underline{\beta}_1$, $\bar{\underline{\beta}}_2$ and σ_{ii} $(i=1,2,\ldots,n)$ respectively when $\Delta_2=0$. Consequently

$$\sup_{\Delta_2=0} L(\underline{\beta}_1,\bar{\underline{\beta}}_2,\underline{\theta}_2|\underline{y},X) = (2\pi)^{-nT/2} \prod_{i=1}^{n} \hat{\sigma}^{*\,-T/2}_{ii}\ e^{-nT/2}\ . \tag{4.10.21}$$

The LR criterion to test the hypothesis $\Delta_2=0$ is given by

$$\hat{\Lambda}_1 = \frac{\sup_{\Delta_2=0} L(\underline{\beta}_1,\bar{\underline{\beta}}_2,\underline{\theta}_2|\underline{y},X)}{\sup L(\underline{\beta}_1,\bar{\underline{\beta}}_2,\underline{\theta}_2|\underline{y},X)}\ . \tag{4.10.22}$$

When the null hypothesis $\Delta_2=0$ is true, the asymptotic distribution of $-2\log_e \hat{\Lambda}_1$ is central χ^2 with $\frac{1}{2}K_2(K_2+1)$ d.f. To simplify our calculations we derive certain estimators which are asymptotically equivalent to ML estimators.

Now we assume

<u>Assumption 4.10.3</u>:

(1) $\lim_{T\to\infty} T^{-1}X_{2i}'X_{2i}$ and $\lim_{T\to\infty} T^{-1}X_{1i}'X_{1i}$ are finite and positive definite matrices.

(2) $\text{Lim } T^{-1}X_{2i}'X_{1i}$ is finite for every i.

(3) Δ_2 is positive definite.

Under Assumption 4.10.3

$$\lim_{T\to\infty} [\Delta_2 + \sigma_{ii}T^{-1}(T^{-1}X_{2i}'X_{2i})^{-1}]^{-1} = \Delta_2^{-1} \quad ,$$

and

$$\lim_{T\to\infty} \left\{\frac{X_{1i}'M_{2i}X_{1i}}{T} + \frac{X_{1i}'X_{2i}}{T}\left(\frac{X_{2i}'X_{2i}}{T}\right)^{-1}\frac{\Delta_2^{-1}}{T}\left(\frac{X_{2i}'X_{2i}}{T}\right)^{-1}\left(\frac{X_{2i}'X_{1i}}{T}\right)\right\}$$

$$= \lim_{T\to\infty}\left\{\frac{X_{1i}'M_{2i}X_{1i}}{T}\right\} \text{ which is finite and nonsingular.}$$

Using these results we can show that the limiting distribution of $\underline{b}_{1i}(\underline{\bar{\beta}}_2,\underline{\theta}_2)$ will be the same as that of

$$\underline{b}_{1i} = (X_{1i}'M_{2i}X_{1i})^{-1}X_{1i}'M_{2i}\underline{y}_i \qquad (i=1,2,\ldots,n) \tag{4.10.23}$$

and the limiting distribution of $\bar{\underline{b}}_2(\underline{\beta}_1,\underline{\theta}_2)$ will be the same as that of

$$\frac{1}{n}\sum_{i=1}^{n}(\hat{\underline{\beta}}_{2i}-\hat{X}_{1i}\underline{b}_{1i}) = \frac{1}{n}\sum_{i=1}^{n}\underline{b}_{2i} \tag{4.10.24}$$

where $\underline{b}_{2i}$ is as defined in (4.10.6) and use is made of the result $(X_{2i}'X_{2i})^{-1}X_{2i}' - (X_{2i}'X_{2i})^{-1}X_{2i}'X_{1i}(X_{1i}'M_{2i}X_{1i})^{-1}X_{1i}'M_{2i} = (X_{2i}'M_{1i}X_{2i})^{-1}X_{2i}'M_{1i}$ which can be easily verified by postmultiplying both sides by X_{1i} and X_{2i}.

Since $\lim_{T\to\infty} \text{tr}[T^{-1}\Delta_2(T^{-1}X_{2i}'X_{2i})^{-1}] = 0$, the limiting distribution of $\tilde{\sigma}_{ii}$ will be the same as

$$\hat{s}_{ii} = \frac{(\underline{y}_i-X_{1i}\underline{b}_{1i})'M_{2i}(\underline{y}_i-X_{1i}\underline{b}_{1i})}{T-K_2}$$

$$= \underline{y}_i'M_i\underline{y}_i/(T-K_2). \tag{4.10.25}$$

The second equality in (4.10.25) follows from the result

$$M_i\underline{y}_i = \underline{y}_i - X_i\underline{b}_i$$

$$= \underline{y}_i - [X_{1i}, X_{2i}]\begin{bmatrix}\underline{b}_{1i}\\(\hat{\underline{\beta}}_{2i}-\hat{X}_{1i}\underline{b}_{1i})\end{bmatrix} \tag{4.10.26}$$

where $X_i = [X_{1i}, X_{2i}]$ and use is made of the result $\underline{b}_i = (X_i'X_i)^{-1}X_i'\underline{y}_i = [\underline{b}_{1i}', (\hat{\underline{\beta}}_{2i} - \hat{X}_{1i}\underline{b}_{1i})']'$, cf. Goldberger (1964, p. 174). We can easily show that the limiting distribution of $\tilde{\Delta}_2$ will be the same as that of

$$n^{-1} S_{b_2} \tag{4.10.27}$$

where S_{b_2} is as defined in (4.10.6).

Substituting $\hat{s}_{ii}$ and $n^{-1}S_{b_2}$ for σ_{ii} and $\Delta_2 + \sigma_{ii}(X_{2i}'X_{2i})^{-1}$ respectively in (4.10.10) we have

$$(2\pi)^{-nT/2} \prod_{i=1}^{n} \left\{ \hat{s}_{ii}^{-\frac{1}{2}(T-K_2)} |X_{2i}'X_{2i}|^{-1/2} |n^{-1}S_{b_2}|^{-\frac{1}{2}} \right\} e^{-nT/2} . \tag{4.10.28}$$

Dividing (4.10.21) by (4.10.28) we have

$$\hat{\Lambda}_1^* = \frac{\prod_{i=1}^{n} \hat{\sigma}_{ii}^{*-T/2}}{\prod_{i=1}^{n} \left\{ \hat{s}_{ii}^{-(1/2)(T-K_2)} |X_{2i}'X_{2i}|^{-1/2} |n^{-1}S_{b_2}|^{-1/2} \right\}} . \tag{4.10.29}$$

The asymptotic distribution of $-2 \log_e \hat{\Lambda}_1^*$ is the same as that of $-2 \log_e \hat{\Lambda}_1$. We can use the asymptotic distribution of $-2 \log_e \hat{\Lambda}_1^*$ to set up a critical region for the test of the hypothesis $\Delta_2 = 0$.

Next we develop an asymptotic procedure to test the hypothesis $\Delta_2 = 0$ and $\underline{\beta}_{11} = \underline{\beta}_{12} = \ldots = \underline{\beta}_{1n} = \bar{\underline{\beta}}_1$ given that $E\underline{\beta}_{2i} = \bar{\underline{\beta}}_2$ for all i.

The likelihood function of parameters of the model (4.10.2a) when $\Delta_2 = 0$ and $\underline{\beta}_{1i} = \bar{\underline{\beta}}_1$, is

$$L(\bar{\underline{\beta}}_1, \bar{\underline{\beta}}_2, \underline{\sigma} | \underline{y}, X) = (2\pi)^{-nT/2} \prod_{i=1}^{n} \sigma_{ii}^{-T/2}$$

$$\cdot \exp \left\{ -\frac{1}{2} \sum_{i=1}^{n} \left[\frac{(\underline{y}_i - X_{1i}\bar{\underline{\beta}}_1 - X_{2i}\bar{\underline{\beta}}_2)'(\underline{y}_i - X_{1i}\bar{\underline{\beta}}_1 - X_{2i}\bar{\underline{\beta}}_2)}{\sigma_{ii}} \right] \right\} . \tag{4.10.30}$$

Sup $L(\bar{\underline{\beta}}_1, \bar{\underline{\beta}}_2, \underline{\sigma} | \underline{y}, X)$ is the same as that in (4.3.66). Taking the ratio of (4.3.66) to (4.10.28) we have

$$\hat{\Lambda}_2^* = \frac{\prod_{i=1}^{n} \hat{\sigma}_{ii}^{-T/2}}{\prod_{i=1}^{n} \left\{ \hat{s}_{ii}^{-(1/2)(T-K_2)} |X_{2i}'X_{2i}|^{-1/2} |n^{-1}S_{b_2}|^{-1/2} \right\}} \quad . \tag{4.10.31}$$

The asymptotic distribution of -2 log$_e \hat{\Lambda}_2^*$ is the same as that of -2 log$_e \hat{\Lambda}_2$, where $\hat{\Lambda}_2$ is the ratio of (4.3.66) to (4.10.17). When the hypothesis $\Delta_2 = 0$ and $\underline{\beta}_{1i} = \bar{\underline{\beta}}_1$ for every i is true, the asymptotic distribution of -2 log$_e \hat{\Lambda}_2$ is central χ^2 with $[\frac{1}{2} K_2(K_2+1) + K_1(n-1)]$ d.f. We can make use of the asymptotic distribution of -2 log$_e \hat{\Lambda}_2^*$ to set up a critical region for the test of the hypothesis $\Delta_2 = 0$ and $\underline{\beta}_{11} = \underline{\beta}_{12} = \ldots = \underline{\beta}_{1n} = \bar{\underline{\beta}}_1$.

Following the LR principle we can also develop an asymptotic test procedure for the hypothesis, $\Delta_2 = 0$ given that $E\underline{\beta}_{2i} = \bar{\underline{\beta}}_2$ for every i and $\underline{\beta}_{1i}$ are the random drawings from the same multivariate distribution. This procedure will be useful in making a choice between the specifications (1.2.4) and (4.2.1).

4.10 Conclusions

We set out the problem of estimating the parameters in a RCR model and mainly achieved asymptotic results. We suggested a consistent and an asymptotically efficient estimator for the mean of a random coefficient vector and an unbiased estimator for its variance-covariance matrix. Under normality assumptions prediction intervals and asymptotic tests of linear hypotheses on the means and variances of random coefficients are constructed. These inference procedures are appropriately modified when the individual disturbances are serially correlated. Further, we studied the optimal forecasting procedures with a RCR model. Finally, we indicated a direction in which some of our basic assumptions can be relaxed. The asymptotic procedure developed in Sections 4.3 and 4.4 can be extended to the case where the disturbances in (4.2.1) are contemporaneously correlated across units, cf. Swamy (1968).

CHAPTER V

A RANDOM COEFFICIENT INVESTMENT MODEL

5.1 Introduction

Grunfeld (1958), in his pioneering work on the determinants of corporate investment, has theorized that the market value of the firm, i.e., the value placed upon the firm by the securities markets, when taken in conjunction with an estimate of the replacement value of the physical assets of the firm, is a sensitive indicator of the expectations upon which corporate investment decisions are based. His analysis of the investment behavior of 11 large U.S. corporations during the years 1935-54 supported his theory. We, in this chapter, assume that the regression coefficient vector appearing in Grunfeld's micro investment function is random across firms following a multivariate distribution, and use the panel data compiled by Grunfeld to illustrate the method developed in Chapter IV.

After summarizing the basic model used by Grunfeld, we connect it with the RCR model in Section 5.2. In Section 5.3 we estimate the means of the random coefficients and study their properties using the analytical results obtained in Chapter IV. In Section 5.4 we develop an aggregate investment function using the results obtained in Section 5.3. We compare fixed coefficient investment models with random coefficient investment models in Sections 5.5 and 5.6. Conclusions of the study are presented in Section 5.7.

5.2 Grunfeld's Hypothesis of Micro Investment Behavior

Grunfeld's investment function involves a firm's current gross investment, y_{ti}, being dependent on the value of the firm's outstanding shares at the beginning of the year $z_{1t-1,i}$, and the firm's beginning-of-year capital stock, $z_{2t-1,i}$.[20] That is, the micro investment function is[21]

$$y_{ti} = \beta_0 + \beta_1 z_{1t-1,i} + \beta_2 z_{2t-1,i} + u_{ti} \quad . \tag{5.2.1}$$

Before we proceed to the analysis of equation (5.2.1) from the standpoint of the RCR model, a few comments on the adequacy of the model (5.2.1) in explaining the micro investment behavior are in order. Grunfeld estimated the equation (5.2.1) including both profit and rate of interest as additional independent variables, but their coefficients turned out to be non-significant. Grunfeld has concluded that profits do not belong in the investment function and that the only reason why they take on a significant positive coefficient in many empirical investment functions is that the proper size and expectations variables have been left out, so that profits serve as a surrogate for those proper variables. The relevant independent variables in a micro investment function are capital stock and value of the firm variables. The former is a size variable and the latter is an expectation variable. According to Grunfeld, producers base their expectations about future profits on the value of the firm variable.

Griliches and Wallace (1965) observed that the failure of Grunfeld's study to uncover a significant interest rate effect could in part be due to his use of annual time series dominated by a period of pegged interest rates. Given the relatively small variance in the observed interest rates, Grunfeld's findings are not surprising. Griliches and Wallace indicated that there was much higher variance in

---

[20]The value of outstanding shares, which is otherwise called the value of the firm, is measured as the total book value of debt at December 31 plus the price of common and preferred shares at December 31 (or average price of December 31 and January 31 of the following year) times the number of common and preferred shares outstanding.

[21]Grunfeld specified a stock adjustment model of the following type:

$$I_t = q_1(z^d_{2t} - z_{2t-1}) + q_2 z_{2t-1}$$

where q_1 is a constant fraction of desired net investment carried out in the year t and q_2 is a constant fraction of the existing capital stock spent for maintenance and repairs. He then hypothesized that the desired stock of capital z^d_{2t} was given by $z^d_{2t} = c_1 + c_2 z_{1t-1}$.

the interest rate variable in the 1948-62 period than was the case in the 1935-54 period examined by Grunfeld. In their study of investment functions using aggregate quarterly data for the period 1948-62, the interest rate variable appeared with a significant coefficient. They concluded that previous output levels and the rate of interest were at least of equal or even greater importance than the market value of the firm variable in determining investment behavior.

In the theory of a firm developed by Jorgenson (1965) the interest rate enters as a variable in the function determining the user cost of capital. The latter variable combined with the value of output forms an explanatory variable for the investment function of a firm.

As shown in footnote 21 the desired capital stock in Grunfeld's theory is a linear function of the value of the firm. In other words, the market value variable is a "proxy" for expected profitability and the relation between expected profits and the market value variable is linear. Gould (1967) has shown that there is no theoretical justification for taking market value or any profit variable as a good indicator of desired capital stock.

Since we are using Grunfeld's data for illustrative purposes we ignore the interest rate variable and other defects of Grunfeld's theory, and consider the model (5.2.1) for the present study.

Assuming that the same equation of the type (5.2.1) is applicable to all corporations[22] in our cross-section sample of size 11, we obtain a system of 11 equations, each of which relates to a single corporation. On each variable included in (5.2.1) we have 20 annual observations for the period 1935-54. These data and their sources have been described in Grunfeld (1958) and also in Boot and de Wit (1960) for 10 corporations excluding American steel foundries.

We can write the set of 11 equations in matrix notation as

$$\underline{y}_i = X_i\underline{\beta}_i + \underline{u}_i \qquad (i=1,2,\ldots,11) \qquad (5.2.2)$$

[22]The names of these corporations are (1) General Motors, (2) U.S. Steel, (3) General Electric, (4) Chrysler, (5) Atlantic Refineries, (6) IBM, (7) Union Oil, (8) Westinghouse, (9) Goodyear, (10) Diamond Match, (11) American Steel Foundries.

where $\underline{y}_i$ is a 20x1 vector of observations on $\underline{y}_t$ for the ith firm, X_i is a 20x3 matrix of observations on $z_{1,t-1}$, z_{2t-1} and $x_{0,t} \equiv 1$ for the ith firm, $\underline{\beta}_i$ is a 3x1 vector of coefficients, $\underline{u}_i$ is a 20x1 vector of disturbances for the ith firm, and $n = 11$. We treat Assumption 4.2.1 as valid for $i,j=1,2,\ldots,11$.

5.3 Estimation and Testing of Random Coefficient Investment Model

We apply the formula (4.3.9) to each equation in (5.2.2) to obtain predictions on each of the 11 $\underline{\beta}_i$'s. Using the formulae (4.3.40) and (4.3.43) we obtain forecast intervals for each firm's coefficient vector. The predictions and their forecast intervals are given in Table 5.1. The results in this Table are obtained by estimating the model (5.2.2) with observations expressed as deviations from their respective firm means -- that is, with an intercept implicitly fitted for each firm. The figures given in Table 5.1 are the results of our computations and our results (with the exclusion of those pertaining to American Steel Foundries) agree with those given in Boot and de Wit (1960). The vector of estimated coefficients $\underline{b}_{zi} = (Z_i' N_T Z_i)^{-1} Z_i' N_T \underline{y}_i$ in Table 5.1 for the ith firm can be viewed as a drawing from a population of $\underline{\beta}_{zi} + (Z_i' N_T Z_i)^{-1} Z_i' N_T \underline{u}_i$ where β_{zi} is a 2x1 vector of slope coefficients, $N_T = I_T - \underline{\iota}_T \underline{\iota}_T' (1/T)$ and Z_i is a Tx2 matrix of observations on $z_{1t-1,i}$ and $z_{2t-1,i}$ for the ith firm. The width of forecast intervals given at the bottom of each estimate can be used for appraising these predictions. The standard error of a prediction is obtained by taking the square root of its estimated variance. The formula for the prediction variance-covariance matrix is $s_{ii}(Z_i' N_T Z_i)^{-1}$.

We treat the set of 11 predictions on slope coefficients given in Table 5.1 as a random sample of size 11 to obtain the sample variance-covariance matrix

$$S_{bz} = \sum_{i=1}^{n} \underline{b}_{zi} \underline{b}_{zi}' - \frac{1}{n} \sum_{i=1}^{n} \underline{b}_{zi} \sum_{i=1}^{n} \underline{b}_{zi}' .$$

$$\frac{S_{bz}}{n-1} = \begin{bmatrix} 0.0029 & -0.0008 \\ & 0.0234 \end{bmatrix} . \tag{5.3.1}$$

Table 5.1

Predictions on Random Coefficients and Their 95% Forecast Intervals for Each Firm[23]

	General Motors	U.S. Steel	General Electric
Coeff. of z_{1t-1}	0.1193 (0.1090) [0.1888]	0.1749 (0.2948) [0.5422]	0.0265 (0.0656) [0.1137]
Standard error	0.0258	0.0742	0.0156
Coeff. of z_{2t-1}	0.3714 (0.1565) [0.2709]	0.3896 (0.6008) [1.0403]	0.1517 (0.1084) [0.1878]
Standard error	0.0371	0.1424	0.0257
Est. Var. Disturbances (s_{ii})	8423.9	9299.6	777.45
	Chrysler	**Atlantic Refineries**	**IBM**
Coeff. of z_{1t-1}	0.0779 (0.0843) [0.1459]	0.1624 (0.2407) [0.4168]	0.1314 (0.1315) [0.2278]
Standard error	0.0200	0.0570	0.0312
Coeff. of z_{2t-1}	0.3157 (0.1216) [0.2106]	0.0031 (0.0926) [0.1606]	0.0854 (0.4233) [0.7330]
Standard error	0.0288	0.0220	0.1003
Est. Var. Disturbances (s_{ii})	176.32	82.17	65.33
	Union Oil	**Westinghouse**	**Goodyear**
Coeff. of z_{1t-1}	0.0875 (0.2770) [0.4795]	0.0529 (0.0662) [0.1148]	0.0754 (0.1433) [0.2481]
Standard error	0.0656	0.0157	0.0339
Coeff. of z_{2t-1}	0.1238 (0.0720) [0.1247]	0.0924 (0.2367) [0.4100]	0.0821 (0.1181) [0.2046]
Standard error	0.0171	0.0561	0.0280
Est. Var. Disturbances (s_{ii})	88.67	104.31	82.79

[23] Figures in ( ) indicate the width of forecast intervals and figures in [ ] indicate the width of simultaneous forecast intervals. The construction of these intervals is strictly valid if z_{1t-1} and z_{2t-1} are nonstochastic. The standard error of a forecast is obtained by dividing the width of the corresponding forecast interval by 2x2.110 where 2.110 is the upper 0.025 point of the t-distribution with 17 d.f.

Table 5.1 (Continued)

	Diamond Match	American Steel Foundries
Coeff. of z_{1t-1}	0.0046 (0.1146) [0.1985]	0.0656 (0.1757) [0.3041]
Standard error	0.0272	0.0416
Coeff. of z_{2t-1}	0.4374 (0.3359) [0.5816]	0.0840 (0.3503) [0.6067]
Standard error	0.0796	0.0830
Est. Var. Disturbances (s_{ii})	1.18	9.83

A consistent estimate of the variance-covariance matrix Δ_z is given by (4.10.6).

$$\frac{1}{n}\sum_{i=1}^{n} s_{ii}(Z_i' N_T Z_i)^{-1} = \begin{bmatrix} 0.0018 & -0.0006 \\ & 0.0047 \end{bmatrix} . \qquad (5.3.2)$$

Subtracting (5.3.2) from (5.3.1) we get

$$\hat{\Delta}_z = \begin{bmatrix} \beta_{1i} & \beta_{2i} \\ 0.0011 & -0.0002 \\ & 0.0187 \end{bmatrix} . \qquad (5.3.3)$$

At this point we digress for a while to continue our discussion in sub-section 4.3(c) of Chapter IV on the possible reasons for getting negative estimates for the variances of random coefficients when we adopt the estimator $\hat{\Delta}$. Firstly, the estimates of the variances of both slope and intercept coefficients in our RCR model came out to be negative when we adopted the estimator (4.10.7) of Δ assuming homoskedasticity of disturbances.

$$\hat{\Delta}^* = \begin{bmatrix} \beta_{0i} & \beta_{1i} & \beta_{2i} \\ -750.81 & 12.25 & 12.27 \\ & -0.14 & -0.05 \\ & & -0.97 \end{bmatrix} . \qquad (5.3.4)$$

On the other hand, when we adopted the estimator (4.3.24) allowing the common variance of disturbances to vary across units, we obtained positive estimates for the variances of slope coefficients. These are given below.

$$\hat{\Delta} = \begin{bmatrix} \beta_{0i} & \beta_{1i} & \beta_{2i} \\ -1012.20 & 0.7263 & -3.5910 \\ & 0.0011 & -0.0002 \\ & & 0.0187 \end{bmatrix} . \qquad (5.3.5)$$

In our present study it turned out that the diagonal elements of $\frac{\hat{\sigma}^2}{n} \cdot \sum_{i=1}^{n} (X_i' X_i)^{-1}$ corresponding to slope coefficients were larger than the corresponding diagonal elements of $\frac{1}{n} \sum_{i=1}^{n} s_{ii}(X_i' X_i)^{-1}$. When we are assuming homoskedasticity we are subtracting larger quantities from some diagonal elements of the variance-covariance matrix, $\frac{S_b}{n-1}$, of estimated regression coefficient vectors, $\underline{b}_i$, than when we are assuming heteroskedasticity. Thus our results in the present example indicate that for any given sample negative estimates for variances might arise with an unbiased estimator of Δ as a result of incorrect assumptions about the form of the error variance-covariance matrix. If we assume homoskedasticity when in fact heteroskedasticity is present we may overestimate some of the diagonal elements of the matrix which we subtract from $\frac{S_b}{n-1}$.

We, further, experimented with the following two-step procedure. We first obtained an estimate of ρ_i by applying nonlinear least squares separately to each of the 11 equations

$$y_{it} = \rho_i y_{it-1} + \beta_{0i}(1-\rho_i) + \beta_{1i}(z_{1t-1,i}-\rho_i z_{1t-2,i}) + \beta_{2i}(z_{2t-1,i} - \rho_i z_{2t-2,i}) + \epsilon_{it} \tag{5.3.6}$$

$$(i=1,2,\ldots,11;\ t=2,3,\ldots,20)$$

and then estimated Δ by

$$\hat{\Delta}^*(\hat{\rho}) = \frac{1}{n-1}\left[\sum_{i=1}^{n} \underline{b}_i^*(\hat{\rho}_i)\underline{b}_i^{*\prime}(\hat{\rho}_i) - \frac{1}{n}\sum_{i=1}^{n} \underline{b}_i^*(\hat{\rho}_i)\sum_{i=1}^{n} \underline{b}_i^{*\prime}(\hat{\rho}_i)\right] - \frac{1}{n}\sum_{i=1}^{n} s_{ii}^*\,(X_i^{*\prime}X_i^*)^{-1} \tag{5.3.7}$$

where $\underline{b}_i^*(\hat{\rho}_i) = (X_i^{*\prime}X_i^*)^{-1}X_i^{*\prime}\underline{y}_i^*$, $X_i^* = \hat{R}_i^* X_i$, $\underline{y}_i^* = \hat{R}_i^*\underline{y}_i$, $s_{ii}^* = \hat{\underline{\epsilon}}_i^{*\prime}\hat{\underline{\epsilon}}_i^*/(T-K+1)$, $\hat{\underline{\epsilon}}_i^* = \underline{y}_i^* - X_i^*\underline{b}_i^*(\hat{\rho}_i)$ and $\hat{R}_i^*$ is obtained from $\hat{R}_i$ in (4.4.4) by deleting the first row.

We can easily show that $\hat{\rho}_i$ and $\hat{\Delta}^*(\hat{\rho})$ are the consistent estimators of ρ_i and Δ respectively. For the present example

$$\hat{\Delta}^*(\hat{\rho}) = \begin{array}{c} \begin{array}{ccc} \beta_{0i} & \beta_{1i} & \beta_{2i} \end{array} \\ \begin{bmatrix} -435.72 & 0.9095 & -0.2439 \\ & 0.0018 & -0.0045 \\ & & 0.0114 \end{bmatrix} \end{array} . \tag{5.3.8}$$

Estimation of Δ under Assumption 4.4.1 clearly resulted in the improved estimates of Δ. In particular, the estimate of the variance of β_{0i} in (5.3.8) is closer to zero than that in (5.3.5).

In order to examine whether Assumption 4.10.2 is appropriate to our data, we have estimated Δ by using the estimator (4.10.9). For the example considered in this chapter we obtained the following values for the elements of (4.10.9).

$$\begin{array}{ccc} \beta_{0i} & \beta_{1i} & \beta_{2i} \end{array}$$
$$\begin{bmatrix} -994.25 & 0.6695 & -3.5975 \\ & 0.0012 & -0.0003 \\ & & 0.0191 \end{bmatrix} . \qquad (5.3.9)$$

Every diagonal element of the above matrix is only slightly bigger than the corresponding diagonal element of the matrix (5.3.5).

An important feature of our computations in the present example is that the sign of the estimated variance of the intercept term shown in (5.3.5) remained negative even when we relaxed the specification (3) in Assumption 4.2.1 in the way shown in Assumptions 4.4.1 and 4.10.2. The signs of the estimated variances of slope coefficients changed from negative to positive when we dropped the homoskedasticity assumption underlying the estimator (4.10.7) and adopted the specification (3) in Assumption 4.2.1. An estimator of Δ depends upon the variation of coefficients across units and the form of the variance-covariance matrix of disturbances. For any well specified model with reasonable assumptions on the form of the variance-covariance matrix of disturbances, an estimator of the variance of a coefficient may assume a negative value due to the sampling fluctuations when there is a low probability for the coefficient to take on a nonzero value or when the true variance of the coefficient is zero. Our computations indicate that the predictions on β_{0i} are not precise and all of them have high standard errors. This may possibly explain why we are obtaining negative estimates for the variance of β_{0i}. Moreover, the intercept terms β_{0i} are affected by the fact that capital stock variable $z_{2t-1,i}$ is measured as a deviation from its value in a year prior to the sample period. We, therefore, decided to retain the specifications (1), (2) and (3) in Assumption 4.2.1 and apply the specifications (4) and (5) to slope coefficients only. We assume further that the intercept term β_{0i} differs nonrandomly across firms. Under these assumptions the development in Section 4.10 is relevant. We can make use of the formulae in Section 4.10 by letting $\underline{\iota}_T$ in (5.2.2) to be the same as X_{1i} in (4.10.1), the matrix of observations on $z_{1t-1,i}$ and $z_{2t-1,i}$ in (5.2.1) for the ith firm to be the same as X_{2i} in (4.10.1), $\underline{\beta}_{0i}$ in (5.2.2) to be the same as $\underline{\beta}_{1i}$ in (4.10.1), the vector of slope coefficients $(\beta_{1i},\beta_{2i})'$ in (5.2.2) to be the same as $\underline{\beta}_{2i}$ in (4.10.1), and $E(\beta_{1i},\beta_{2i})' = (\bar{\beta}_1,\bar{\beta}_2)'$ in (5.2.2) to be the same as $E\underline{\beta}_{2i} = \bar{\underline{\beta}}_2$ in (4.10.1). We first express each observation on $z_{1t-1,i}$, $z_{2t-1,i}$ and y_{ti} as a deviation from the respective firm mean and then estimate the model.

Using the estimates given by $\hat{\Delta}_z$ and s_{ii} we can calculate the standard errors of the estimates of $\bar{\underline{\beta}}_z$ which is the expectation of the vector of slope coefficients in (5.2.2). An estimate of the variance-covariance matrix of $\underline{b}_{zi}$, when it is regarded as an estimate of $\bar{\underline{\beta}}_z$, is given by

$$\hat{\Delta}_z + s_{ii}(Z_i'N_TZ_i)^{-1} \quad . \tag{5.3.10}$$

The positive square root of a diagonal element of (5.3.10) will give the standard error of an estimate of the corresponding element of $\bar{\underline{\beta}}_z$. Estimates of $\bar{\underline{\beta}}_z$ and their standard errors are presented in Table 5.2 below. The standard errors of estimates in Table 5.2 are obviously larger than those of predictions in Table 5.1. The statistic $\underline{b}_{zi}$ is not as precise for estimating $\bar{\underline{\beta}}_z$ as it is for predicting $\underline{\beta}_{zi}$.

Due to the specification (3) in Assumption 4.10.1 all the $\underline{b}_{zi}$'s estimate the same parametric vector $\bar{\underline{\beta}}_z$. We can combine the $\underline{b}_{zi}$'s so as to get a single, consistent and asymptotically efficient estimator for $\bar{\underline{\beta}}_z$. Such a pooled estimator is obtained by replacing $\underline{\theta}_2$ by $\hat{\underline{\theta}}_2$ in (4.10.4).

$$\bar{\underline{b}}_z(\hat{\underline{\theta}}_2) = \begin{bmatrix} \bar{b}_1(\hat{\underline{\theta}}_2) \\ \\ \bar{b}_2(\hat{\underline{\theta}}_2) \end{bmatrix} = \begin{bmatrix} 0.0741 \\ (0.0104) \\ \\ 0.1946 \\ (0.0412) \end{bmatrix} . \tag{5.3.11}$$

The figures given in parentheses are the estimated asymptotic standard errors of $\bar{b}_1(\hat{\underline{\theta}}_2)$ and $\bar{b}_2(\hat{\underline{\theta}}_2)$. They are obtained by taking the square root of the diagonal elements of $\frac{\hat{\Delta}_z}{n}$ where $\hat{\Delta}_z$ is as shown in (4.10.6).

Table 5.2

Consistent Estimates of Means of Random Coefficients and Their Standard Errors for each Firm

Mean of the	General Motors	U.S. Steel	General Electric
Coeff. of z_{1t-1}	0.1193 (0.0424)	0.1749 (0.0812)	0.0265 (0.0360)
Coeff. of z_{2t-1}	0.3714 (0.1417)	0.3896 (0.1974)	0.1517 (0.1393)
Mean of the	Chrysler	Atlantic Refineries	IBM
Coeff. of z_{1t-1}	0.0779 (0.0387)	0.1624 (0.0655)	0.1314 (0.0458)
Coeff. of z_{2t-1}	0.3157 (0.1397)	0.0031 (0.1386)	0.0854 (0.1697)
Mean of the	Union Oil	Westinghouse	Goodyear
Coeff. of z_{1t-1}	0.0875 (0.0735)	0.0529 (0.0360)	0.0754 (0.0469)
Coeff. of z_{2t-1}	0.1238 (0.1379)	0.0924 (0.1476)	0.0821 (0.1397)
Mean of the	Diamond Match	American Steel Foundries	
Coeff. of z_{1t-1}	0.0046 (0.0424)	0.0656 (0.0529)	
Coeff. of z_{2t-1}	0.4374 (0.1581)	0.0840 (0.1600)	

In order to study how widely the values of each slope coefficient in (5.2.1) are spread across firms on either side of their mean value, we furnish below a measure called the coefficient of variation. The coefficient of variation is calculated as $(\hat{\sigma}_{\beta_j}/\bar{b}_j(\hat{\underline{\theta}}_2)) \cdot 100(j=1,2)$, where σ_{β_j} is the standard deviation of β_j.

$$\begin{array}{ll} & \text{Estimate of the} \\ \text{Coeff. of var. of } \beta_{1i} & 45.07\% \\ \text{Coeff. of var. of } \beta_{2i} & 69.89\%. \end{array} \qquad (5.3.12)$$

As indicated by the above values of the coefficients of variation there is a substantial variation in coefficients across firms. In the light of this result, it is hard to believe that the slope coefficients in (5.2.2) are the same for all firms.

The reduced form coefficients β_{1i} and β_{2i} are related to the structural coefficients q_{1i}, q_{2i} are c_{2i} in the following manner:

$$\beta_{1i} = q_{1i}c_{2i} \text{ and } \beta_{2i} = q_{2i} - q_{1i} \quad . \qquad (5.3.13)$$

The nature of these structural coefficients is explained in footnote 21. We know a priori that the values of coefficients q_{1i} and q_{2i} lie between 0 and 1. Grunfeld has conjectured that $0 < q_{2i} < 0.3$ for each i. Since β_{2i} is linearly related to q_{1i} and q_{2i},

$$E\beta_{2i} = Eq_{2i} - Eq_{1i} \quad . \qquad (5.3.14)$$

Knowing a priori the value of Eq_{2i} we can estimate Eq_{1i} since we have an estimate of $E\beta_{2i}$ in (5.3.11). If q_{1i} and c_{2i} are uncorrelated, then

$$E\beta_{1i} = Eq_{1i}Ec_{2i} \quad . \qquad (5.3.15)$$

From this equation we can estimate the mean of c_{2i} using the estimate of the mean of β_{1i} given in (5.3.11) and an estimate of the mean of q_{1i} given by (5.3.14).

If q_{1i} and c_{2i} are correlated,[24] then from the definition of covariance it follows that

$$E\beta_{1i} = Eq_{1i}c_{2i} = \text{cov}(q_{1i}, c_{2i}) + Eq_{1i}Ec_{2i} \quad . \tag{5.3.16}$$

This clearly points out that even if we know $E\beta_{1i}$ and Eq_{1i} we cannot say anything about the value of Ec_{2i} unless we know the covariance of c_{2i} and q_{1i}. In general, we will not have prior information on the covariance of c_{2i} and q_{1i}.

Next question is: what can we say about the variances of structural coefficients if we have the estimates of the variances and covariances of reduced form coefficients? We have from (5.3.13) that

$$\sigma^2_{\beta_2} = \sigma^2_{q_2} + \sigma^2_{q_1} - 2\ \text{cov}(q_{1i}, q_{2i}) \tag{5.3.17}$$

where $\sigma^2_{\beta_2}$, $\sigma^2_{q_2}$, $\sigma^2_{q_1}$ stand for the variances of β_{2i}, q_{2i} and q_{1i} respectively. Since the estimate of $\sigma^2_{\beta_2}$ is given in (5.3.3), we can estimate σ^2_{q1} knowing a priori the value of $\sigma^2_{q_2}$ and $\text{cov}(q_{1i}, q_{2i})$. When q_{1i} and c_{2i} are independent, we know from (5.3.13) that

$$\sigma^2_{\beta_1} = (\sigma^2_{q_1} + \bar{q}^2_1)(\sigma^2_{c_2} + \bar{c}^2_2) - \bar{q}^2_1 \bar{c}^2_1 \tag{5.3.18}$$

where $\bar{q}_1 = Eq_{1i}$, $\bar{c}_2 = Ec_{2i}$, and $\sigma^2_{\beta_1}$, $\sigma^2_{c_2}$ are the variances of β_{1i} and c_{2i} respectively. If we have the estimates of all but $\sigma^2_{c_2}$, we can estimate $\sigma^2_{c_2}$ from (5.3.18). If q_{1i} and c_{2i} are correlated, the expression for the variance of $q_{1i}c_{2i}$ is slightly long, cf. Goodman (1962). If q_{1i} and c_{2i} follow bivariate normal with finite means, variances and covariance, following Craig (1936, p. 9) we can show that

$$\sigma^2_{\beta_1} = \bar{q}^2_1\sigma^2_{c_2} + \bar{c}^2_2\sigma^2_{q_1} + 2\bar{q}_1\bar{c}_2\ \text{cov}(q_1, c_2) + (1+\rho^2)\sigma^2_{q_1}\sigma^2_{c_2} \tag{5.3.19}$$

where ρ is the correlation coefficient of q_{1i} and c_{2i}. We need prior information about ρ if we want to estimate $\sigma^2_{c_2}$ given the estimates of $\bar{q}_1$, $\bar{c}_2$ and $\sigma^2_{\beta_1}$. The

[24] The distribution of the product of two correlated normal variables is studied by Craig (1936, 1942).

equations (5.3.14)-(5.3.18) are valid whatever be the form of the distribution of structural or reduced form coefficients provided they have finite second order moments.

In order to make use of the testing procedures developed in Chapter IV, we make Assumption 4.3.1.[25] The estimates in (5.3.11) indicate that the means of β_{1i} and β_{2i} are significantly different from zero. On the basis of the estimated means and variances of β_{1i} and β_{2i} we can approximately evaluate the probability that the normally distributed variables β_{1i} and β_{2i} take values within the intervals $\bar{\beta}_1 \pm 2\sigma_{\beta_1}$ and $\bar{\beta}_2 \pm 2\sigma_{\beta_2}$ respectively. Here σ_{β_1} is the standard deviation of β_{1i} and σ_{β_2} is the standard deviation of β_{2i}. Thus

$$\Pr(0.007 < \beta_{1i} < 0.141) = 0.95 \qquad (5.3.20a)$$

and

$$\Pr(-0.077 < \beta_{2i} < 0.467) = 0.95 \quad . \qquad (5.3.20b)$$

These interval statements hold only approximately because they are based on estimated means and standard deviations.

Using the statistic in (4.3.60) we test the hypothesis that $\underline{\bar{\beta}}_z = 0$. For the present example the value of (4.3.60) under the above hypothesis is equal to 30.87 and this falls well above the 5% value of F for 2, 9 d.f., so that the data cannot be regarded as consistent with the hypothesis.

---

[25] Some of the coefficients in Grunfeld's model are thought to be positive lying between zero and one. Therefore the normality assumption is not entirely appropriate here. Moreover this assumption has some implications for the distributions of structural coefficients. If β_{1i} and β_{2i} follow bivariate normal, c_{2i} will be distributed as the ratio of two correlated normal variables which does not have any finite moments, cf. [Fieller (1932), Craig (1942), and Marsaglia (1965)]. Therefore the assumption that β_{1i} and β_{2i} are jointly normal is a contradiction to our basic assumption that c_{2i} has a finite mean and variance. In order to avoid such contradictions we redefine $\bar{c}$ as the median of c_{2i} and $\sigma^2_{c_2}$ as the mean deviation or some other measure of dispersion which is finite. We also assume that the probability of values of β_{1i} and β_{2i} outside their relevant ranges is very small.

Using the asymptotic distribution of $\underline{\ell}'\hat{\Delta}_z\underline{\ell}$, which is shown in (4.3.61) we obtain the following 95% confidence intervals for the variances of the marginal distributions of β_{1i} and β_{2i}.

$$0.0005 < \sigma^2_{\beta_1} < 0.0034 \tag{5.3.21a}$$

$$0.0091 < \sigma^2_{\beta_2} < 0.0575. \tag{5.3.21b}$$

If the above interval estimates, based on a single sample, cover the true values of parameters, then they indicate that the value of $\sigma^2_{\beta_1}$ is small lying between 0.0005 and 0.0034.

5.4 Aggregate Investment Function

Premultiplying (5.2.2) by N_T in (2.3.7) and writing $\underline{\beta}_{zi} = \underline{\bar{\beta}}_z + \underline{\delta}_{zi}$ we have

$$N_T\underline{y}_i = N_T Z_i \underline{\bar{\beta}}_z + N_T Z_i \underline{\delta}_{zi} + N_T \underline{u}_i \qquad (i=1,2,\ldots,11) \quad . \tag{5.4.1}$$

Aggregating (5.4.1) over i we have

$$N_T\underline{\bar{y}} = N_T \bar{Z}\, \underline{\bar{\beta}}_z + N_T \sum_{i=1}^{n} Z_i \underline{\delta}_{zi} + N_T \underline{\bar{u}} \tag{5.4.2}$$

where $\underline{\bar{y}} = \sum_{i=1}^{n} \underline{y}_i$, $\bar{Z} = \sum_{i=1}^{n} Z_i$ and $\underline{\bar{u}} = \sum_{i=1}^{n} \underline{u}_i$.

From (5.3.11) we have

$$\bar{y}^*_t = \underset{(0.0104)}{0.0741}\, \bar{z}^*_{1t-1} + \underset{(0.0412)}{0.1946}\, \bar{z}^*_{2t-1} + \sum_{i=1}^{n} \hat{\delta}_{1i} \bar{z}^*_{1t-1,i} + \sum_{i=1}^{n} \hat{\delta}_{2i} \bar{z}^*_{2t-1,i} + \hat{u}_t \ , \tag{5.4.3}$$

or ignoring the error terms

$$\hat{\bar{y}}^*_t = \underset{(0.0104)}{0.0741}\ \bar{z}^*_{1t-1} + \underset{(0.0412)}{0.1946}\ \bar{z}^*_{2t-1} \tag{5.4.4}$$

where $\bar{y}^*_t$, $\bar{z}^*_{1t-1}$, and $\bar{z}^*_{2t-1}$ are the elements of $N_T\underline{\bar{y}}$, $\sum_{i=1}^{n} N_T\underline{z}_{1i}$ and $\sum_{i=1}^{n} N_T\underline{z}_{2i}$ respectively, $\underline{z}_{1i}$ is a Tx1 vector of observations on $z_{1t-1,i}$ and $\underline{z}_{2i}$ is a Tx1 vector of observations on $z_{2t-1,i}$.

The above equation should not be confused with an aggregate equation (4.5.2) derived in Section 4.5 of Chapter IV. What is shown in Section 4.5 is how to estimate an aggregate equation using the aggregated data. The equation (5.4.4) is obtained by simply aggregating the estimated micro-relations (5.2.2).

The equation (5.4.4) indicates that at the aggregate level the value of the firm variable z_{1t-1} has a significant effect on the investment. From this we conclude that within the framework of the model (5.2.1) the market value variable is important in the aggregate investment function.

5.5 Comparison of Random Coefficient Model with Fixed Coefficient Macro Model

Boot and de Wit (1960) have estimated an aggregate investment function of the type (5.4.2) using the aggregated data for 10 corporations excluding American Steel Foundries. Their results are as follows:

$$\bar{y}^*_t = \underset{(0.025)}{0.099}\ \bar{z}^*_{1t-1} + \underset{(0.020)}{0.260}\ \bar{z}^*_{2t-1} + \hat{u}_t \qquad (5.5.1)$$

where $\bar{y}^*_t = \sum_{i=1}^{n} y_{ti} - \frac{1}{T}\sum_{t=1}^{T}\sum_{i=1}^{n} y_{ti}$, $\bar{z}^*_{1t-1} = \sum_{i=1}^{n} z_{1t-1,i} - \frac{1}{T}\sum_{t=1}^{T}\sum_{i=1}^{n} z_{1t-1,i}$, $\bar{z}^*_{2t-1} = \sum_{i=1}^{n} z_{2t-1,i} - \frac{1}{T}\sum_{t=1}^{T}\sum_{i=1}^{n} z_{2t-1,i}$, T=20 and n=10. The underlying assumption of the model (5.5.1) is that the coefficient vector is fixed (or nonrandom) and the same for all corporations.

If the regression coefficient vectors $\underline{\beta}_i$ are fixed but different for different firms, then a simple aggregation of (5.2.2) across units introduces biases into coefficient estimators, unless the covariances between the micro coefficients and the X_i's in (1.2.25) are zero. The mathematical expectation of a macro coefficient estimator will, in general, depend on a complicated combination of corresponding and noncorresponding micro coefficients. The aggregation biases contained in the macro estimates in (5.5.1) have been evaluated by Boot and de Wit under the assumption that $\underline{b}_i = \underline{\beta}_i$ with probability one and $\sigma_{11} = \sigma_{22} = \dots = \sigma_{nn}$. The standard errors of estimates in (5.5.1) are also affected by the aggregation bias and "implied standard error" in $\hat{u}_t$, cf. Boot and de Wit (1960).

If the coefficient vector $\underline{\beta}_i$ is random across units, then as shown in Section 4.5, the macro estimates in (5.5.1) are inefficient and the standard errors in (5.5.1) are obtained by utilizing an inappropriate formula.

5.6 Comparison of Random Coefficient Model with Fixed Coefficient Micro Model

Let us assume for purposes of this section that the vector of slope coefficients in (5.2.2) is nonrandom and the same for all individual units. That is,

$$\underline{\beta}_{z1} = \underline{\beta}_{z2} = \ldots = \underline{\beta}_{z11} = \underline{\beta}_z, \text{ say }. \tag{5.6.1}$$

Under this restriction we can write (5.2.2) as

$$\begin{bmatrix} \underline{y}_1 \\ \underline{y}_2 \\ \cdot \\ \cdot \\ \cdot \\ \underline{y}_{11} \end{bmatrix} = N_T \begin{bmatrix} Z_1 \\ Z_2 \\ \cdot \\ \cdot \\ \cdot \\ Z_{11} \end{bmatrix} \underline{\beta}_z + N_T \begin{bmatrix} \underline{u}_1 \\ \underline{u}_2 \\ \cdot \\ \cdot \\ \cdot \\ \underline{u}_{11} \end{bmatrix} . \tag{5.6.2}$$

If the specification (3) in Assumption 4.2.1 is true, the BLUE of $\underline{\beta}_z$, à la Aitken, is

$$\hat{\underline{\beta}}_z(\underline{\sigma}) = \left[\sum_{j=1}^{n} \frac{Z_j' N_T Z_j}{\sigma_{jj}}\right]^{-1} \left[\sum_{i=1}^{n} \frac{Z_i' N_T \underline{y}_i}{\sigma_{ii}}\right] . \tag{5.6.3}$$

Since the σ_{ii}'s are unknown, we can obtain a consistent and an asymptotically efficient estimator of $\underline{\beta}_z$ by substituting for the σ_{ii}'s their unbiased estimators in (4.3.21), cf. Zellner (1962a).[26]

$$\hat{\underline{\beta}}_z(\underline{s}) = \left[\sum_{i=1}^{n} \frac{Z_i' N_T Z_i}{s_{ii}}\right]^{-1} \left[\sum_{i=1}^{n} \frac{Z_i' N_T \underline{y}_i}{s_{ii}}\right] . \tag{5.6.4}$$

---

[26] When the X_i's are nonstochastic the estimator (5.6.4) is in fact an unbiased estimator of $\underline{\beta}$ under the symmetry assumption about the distribution of disturbances, cf. Kakwani (1967).

We can also easily show that the estimated asymptotic variance-covariance matrix of $\hat{\underline{\beta}}_z(s)$ is given by

$$\left[\sum_{i=1}^{n} \frac{Z_i' N_T Z_i}{s_{ii}}\right]^{-1} . \qquad (5.6.5)$$

For our investment model using again the observations expressed as deviations from their respective firm means we obtained the following values of $\underline{\beta}_z$.

$$\hat{\underline{\beta}}_z(s) = \begin{bmatrix} \hat{\beta}_1(s) \\ \\ \hat{\beta}_2(s) \end{bmatrix} = \begin{bmatrix} 0.0821 \\ (0.0057) \\ 0.1126 \\ (0.0077) \end{bmatrix} . \qquad (5.6.6)$$

The figures in parentheses are the square roots of the diagonal elements of (5.6.5). According to the results in (5.6.6) both the coefficients are significant.

Because of the restriction (5.6.1) we can immediately write down the aggregate investment function as

$$\sum_{i=1}^{n} N_T \underline{y}_i = \sum_{i=1}^{n} N_T Z_i \underline{\beta}_z + \sum_{i=1}^{n} N_T \underline{u}_i ; N_T \underline{\bar{y}} = N_T \bar{Z} \underline{\beta}_z + N_T \underline{\bar{u}} . \qquad (5.6.7)$$

Substituting $\hat{\underline{\beta}}_z(s)$ in (5.6.6) for $\underline{\beta}_z$ gives

$$\bar{y}_t^* = \underset{(0.0057)}{0.0821} \bar{z}_{1t-1}^* + \underset{(0.0077)}{0.1126} \bar{z}_{2t-1}^* + \hat{u}_t . \qquad (5.6.8)$$

The above estimates are more efficient than those in (5.5.1) if the assumption of heteroskedasticity of disturbances across firms is correct. The assumption of homoskedasticity appears to be unrealistic since the estimates of error variances in Table 5.1 show substantial variation across our sample firms.

The differences between the estimates (5.6.8) and the estimates (5.4.4) are mainly due to differences in their underlying assumptions. The conclusion that emerges from the comparison of the results in (5.4.4), (5.5.1), and (5.6.8) is that the estimates of coefficients in (5.2.2) are not robust to significant departures in the assumption of constancy of regression coefficients across firms. Now the question is: which of these assumptions are valid for the situation under study? The values of the coefficients of variation in (5.3.12) indicate that the coefficient vector in (5.2.2) varies substantially across firms. Now we will test the assumption (5.6.1) using Zellner's statistic in (4.3.80). If the assumption (5.6.1) is true, then $\underline{b}_{zi}$ $(i=1,2,\ldots,n)$ are the unbiased, consistent, and mutually uncorrelated estimators of the same parametric vector $\underline{\beta}_z$. Let the distribution of $(\underline{b}_{zi} - \underline{\beta}_z)$ be $N_{k'}(0,\sigma_{ii}(Z_i'N_TZ_i)^{-1})$. Consider the homogeneity statistic

$$\frac{1}{K'(n-1)} \sum_{i=1}^{n} \frac{[\underline{b}_{zi}-\hat{\underline{\beta}}_z(\underline{s})]'Z_i'N_TZ_i\,[\underline{b}_{zi}-\hat{\underline{\beta}}_z(\underline{s})]}{s_{ii}} \tag{5.6.9}$$

where

$$\hat{\underline{\beta}}_z(\underline{s}) = \left[\sum_{i=1}^{n} \frac{Z_j'N_TZ_j}{s_{jj}}\right]^{-1} \left[\sum_{i=1}^{n} \frac{Z_i'N_TZ_i}{s_{ii}}\underline{b}_{zi}\right] \text{ and } K' = K-1.$$

If all the $\underline{b}_{zi}$'s do not estimate the same parametric vector, then the differences are reflected in the statistic (5.6.9) and, therefore, it may be used to test (5.6.1). The asymptotic distribution of (5.6.9) is F with $(n-1)K'$, $n(T-K)$ d.f. as $T\to\infty$ under the hypothesis that $\underline{\beta}_{zi}=\underline{\beta}_z$ for all i, and the assumptions that the s_{ii}'s are continuous functions and n is fixed.

Using the values given in Table 5.1 and in (5.6.6), we obtain the value 14.45 for (5.6.9) and this falls well above the 5% value of F with 20,187 d.f. so that the data do not support the hypothesis that the coefficient vectors of different firms are all equal.

Since the condition (5.6.1) is not supported by our data, the results in (5.5.1) contain aggregation biases and the results in (5.6.8) suffer from a specification error indicated by Zellner (1962b). The estimates in (5.3.13) are not based on the condition (5.6.1). We may be justified in allowing the slope coefficients to vary randomly across firms.

In order to test for the randomness of slope coefficients we test the hypothesis

$$H_0:\ \Delta_z = 0 \text{ given that } E\underline{\beta}_{zi} = \bar{\underline{\beta}}_z \text{ for all } i \tag{5.6.10}$$

using the statistic (4.10.29). For the present example

$$T \sum_{i=1}^{n} \log_e \hat{\sigma}_{ii} = 1216.3540, \quad (T-2) \sum_{i=1}^{n} \log_e s_{ii} = 928.0402,$$

$$\sum_{i=1}^{n} \log_e |Z_i' Z_i| = 258.5525, \text{ and } n \log_e |n^{-1} S_{bz}| = -107.0168. \quad (5.6.11)$$

The iterative procedure used in solving (4.10.19) and (4.10.20) is stopped when there was no change in the fifth decimal place. From the values in (5.6.11) we have $-2 \log_e \hat{\Lambda}_c^* = 136.7781$ which is well above the chi-square value with 3 d.f. The data do not support the hypothesis that the variance-covariance matrix of slope coefficients in (5.2.2) is null. This indicates that for the situation under study it may be appropriate to treat the slope coefficients as random. The outcome of the above test also indicates that the matrix (4.3.16) is significantly different from null and the OLS estimator of $\underline{\bar{\beta}}_z$ cannot be more efficient than $\underline{\bar{b}}_z(\underline{\hat{\theta}})$ even in small samples.

5.7 Conclusions

We have derived an aggregate investment function for a group of 11 major corporations in the U.S. from Grunfeld's micro investment function making a general assumption that the vector of slope coefficients is random across firms but distributed with the same mean and the same variance-covariance matrix. We have treated the intercept term as fixed and different for different firms. Our empirical results based on the same body of data as was analyzed by Grunfeld indicate that the mean of the coefficient of the value of the firm variable is significantly different from zero. Our statistical tests indicate that we cannot treat the coefficient vector of the investment function as fixed and the same for all firms. The coefficient vector exhibits substantial variation across firms.

CHAPTER VI

AGGREGATE CONSUMPTION FUNCTION WITH COEFFICIENTS RANDOM ACROSS COUNTRIES

6.1 Introduction

The problem of estimating an economic relationship from the combined time series and cross-country data poses a serious challenge to econometricians, cf. Houthakker (1961, 1962, 1965). Houthakker (1961, 1962) estimated an aggregate inter-country savings function to find out the determinants of savings in developed and underdeveloped countries. He computed, among other things, a weighted regression of per capita personal savings on per capita disposable income with weights inversely proportional to the square of personal income per capita. These weights were determined on an *ad hoc* basis. Such studies, useful as they are, suffer from serious limitations imposed by the procedure of pooling data on countries with different levels of development and economic activity. A simple weighting scheme like the one adopted by Houthakker may not eliminate all the heterogeneity among countries. To understand the limitations of these studies it is always desirable to subject the same body of data to different types of analyses based on different sets of assumptions. Zellner and Sankar (1967) analyzed Houthakker's data from the standpoint of errors-in-variables model in an effort to determine whether the average propensity to consume depends on the level of income. In this chapter we make an attempt to analyze cross-country data on aggregate consumption expenditures starting from a different set of assumptions. We first set up an aggregate consumption model on the basis of some existing theories of consumption and assume that such a function is defined for each country appearing in our cross-section sample. Then we test whether the coefficient vector of that model is the same for all countries. If it is not the same, then we have a problem. If the coefficient vector instead of being fixed is distributed randomly across countries with the same mean, then we have a solution to the problem. If this assumption is true, then the data on all countries contain information on the same parametric vector (i.e., the mean of a coefficient vector) and they can be pooled in some way to estimate the common mean.

In Section 6.2 we specify the model and in Section 6.3 we indicate the source and nature of data used in this study. In Section 6.4 we discuss the results obtained in a pooled regression in which the parameters are constrained to be the same for all countries and test whether the countries under study are homogeneous in so far as the regression coefficient vectors are concerned. In Section 6.5 we discuss the results of a random coefficient approach. We furnish conclusions of this study in Section 6.6.

6.2 Aggregate Consumption Model

We primarily experiment with two types of consumption models in this study. One is based on the simple Keynesian hypothesis and the other is based on the habit persistence hypothesis putforth by Brown (1952).

Model I:

$$C_{it} = \alpha_{0i} + \alpha_{1i} x_{it} + v_{it} \quad (i=1,2,\ldots,n;\ t=1,2,\ldots,T) \qquad (6.2.1)$$

Model II:

$$C_{it} = \gamma_{1i} C_{i,t-1} + \gamma_{2i} x_{it} + \varepsilon_{it} \quad (i=1,2,\ldots,n;\ t=1,2,\ldots,T)\ . \qquad (6.2.2)$$

The subscript i denotes an observation for the ith country and the subscript t denotes an observation for the t-th year. In Model I we relate consumption to measured income and in Model II we relate consumption to lagged consumption and measured income.[27] As is well known, the latter formulation cannot be distinguished from the formulation based on the permanent income hypothesis put forward by Friedman (1957).[28] The permanent income hypothesis leads to a complicated reduced

---

[27] Brown in his original formulation of habit persistence hypothesis did not omit the intercept term. Since we had data for only eight years, when we fitted the model (6.2.2) with a constant term we ran into multicollinearity problems.

[28] According to Friedman, $C_{it} = k_i x^p_{it} + e_{it}$ where e_{it} denotes the transitory consumption with some stochastic properties and x^p_{it} denotes the i-th country's permanent income in the t-th year. If we take $x^p_{it} = (1-\lambda_i) \sum_{j=1}^{\infty} \lambda_i^j x_{it-j}$ and apply Koyck-transformation we obtain $C_{it} = \lambda_i C_{it-1} + k_i(1-\lambda_i) x_{it} + e_{it} - \lambda_i e_{it-1}$ whose systematic part is exactly like that of (6.2.2). However, the presence of a structural parameter in the error term can be utilized to distinguish Friedman's model from (6.2.2), cf. Zellner and Geisel (1968).

form equation which cannot be estimated with the type of data (to be explained below) at our disposal, cf. Zellner et al. (1965). We only try to interpret our results from the standpoint of permanent income hypothesis. In this study we abstract from possible "simultaneous equation" complications.

6.3 Source and Nature of Data

The source of data for our study is the U.N. Yearbook of National Accounts Statistics, 1965. From the table relating to "Receipts and Expenditures of Households and Non-Profit Institutions", we collected data on consumption expenditures and disposable income relating to 24 countries.[29] These figures in current prices have been deflated by cost-of-living indices with 1958 = 100 published in U.S. Statistical Year Book, 1965. The figures in local currency at constant prices have been converted into constant U.S. dollars by using parity exchange rates for the year 1958 published in U.N. National Accounts Statistics, 1965. Using the population figures given in Demographic Year Book, 1965, we obtained per capita real consumption, C_{ti} and per capita real disposable income, x_{ti} in U.S. 1958 dollars for each country for the period 1955-1963.

6.4 Fixed Coefficient Approach

For convenience we relabel the variables as follows:

$$\underline{y}_i = X_i \underline{\beta}_i + \underline{u}_i \qquad (i=1,2,\ldots,n) \qquad (6.4.1)$$

where $\underline{y}_i$, X_i, $\underline{\beta}_i$ and $\underline{u}_i$ are as explained in (4.2.1). The equation (6.4.1) represents a consumption function for the i the country. When (6.4.1) represents the Model I, K is equal to 2, the first column of X_i is a TX1 vector of unit elements,

[29] The countries are (1) Austria, (2) Australia, (3) Belgium, (4) Burma, (5) Canada, (6) Chile, (7) Colombia, (8) Costa Rica, (9) Denmark, (10) Ecuador, (11) Finland, (12) Honduras, (13) Ireland, (14) Jamaica, (15) Japan, (16) Mauritius, (17) Netherlands, (18) New Zealand, (19) Spain, (20) South Africa, (21) Sweden, (22) Switzerland, (23) U.K., (24) Taiwan. While the U.N. Year Book contains national accounts data for all the member nations of the U.N., for only 24 countries does it provide the required data on a comparable basis for more than a handful of years.

and the second column of X_i is a TX1 vector of time series observations on x_{ti}. The coefficient vector $\underline{\beta}_i$ contains α_{0i} and α_{1i} as its elements, and $\underline{u}_i = (v_{i1},\ldots,v_{iT})'$. When the equation (6.4.1) represents the Model II, K is equal to 2, the first column of X_i is a TX1 vector of the time series observations on C_{it-1} and the second column is a TX1 vector of time series observations on x_{it}. The coefficient vector $\underline{\beta}_i$ contains γ_{1i} and γ_{2i} as its elements, and $\underline{u}_i \equiv (\epsilon_{i1},\ldots,\epsilon_{iT})'$.

We first estimate the Models I and II under the following assumption for i and j=1,2,...,n.

Assumption 6.4.1:

(1) $\underline{\beta}_1 = \underline{\beta}_2 = \ldots = \underline{\beta}_n = \underline{\beta}$.

(2) $E\underline{u}_i = 0$ and $E\underline{u}_i\underline{u}_j' = \begin{cases} \sigma^2 I & \text{if } i = j \text{ ;} \\ 0 & \text{otherwise .} \end{cases}$

(3) The variable $C_{i,t-1}$ is uncorrelated with ϵ_{it} and the variable x_{it} is uncorrelated with both v_{it} and ϵ_{it}.

The above assumptions describe the full ideal conditions under which the Models I and II can be estimated by the classical least squares method. By pooling the time series data from all the 24 countries we obtain 192 observations on each variable. Using these observations and the formula

$$\underline{b} = \left[\sum_{j=1}^{n} X_j'X_j\right]^{-1} \left[\sum_{i=1}^{n} X_i'X_i\underline{b}_i\right] \tag{6.4.2}$$

where $\underline{b}_i = (X_i'X_i)^{-1}X_i'\underline{y}_i$, we obtain the following estimates for the coefficients of Models I and II.

OLS Estimates[30]

Model I:
$$C_{it} = \underset{(3.071)}{10.538} + \underset{(0.004)}{0.893}\ x_{it} \tag{6.4.3}$$

Model II:
$$C_{it} = \underset{(0.033)}{0.599}\ C_{i,t-1} + \underset{(0.029)}{0.375}\ x_{it} \ . \tag{6.4.4}$$

---

[30] We omitted the intercept term from Model II to facilitate comparison with results reported later in this chapter.

The figures given in parentheses are the standard errors calculated by taking the square root of the diagonal elements of $\hat{\sigma}^2 \left[\sum_{i=1}^{n} X_i' X_i \right]^{-1}$ where $\hat{\sigma}^2 = \underline{y}' M \underline{y} / (nT-K)$, $\underline{y}$ and X are as explained in (4.3.2a) and $M = I - X(X'X)^{-1}X'$. If Assumption 6.4.1 is correct, the above estimates are consistent and asymptotically efficient. It is likely that the specification (2) in Assumption 6.4.1 is not valid for our data. We relax it in the following manner.

Assumption 6.4.2:

(1) Same as (1) in Assumption 6.4.1.

(2) $E\underline{u}_i = 0$ and $E\underline{u}_i\underline{u}_j' = \begin{cases} \sigma_{ii} I & \text{if } i=j; \\ 0 & \text{otherwise} . \end{cases}$

(3) Same as (3) in Assumption 6.4.1.

It is easy to see what this assumption means. The estimator which yields consistent and asymptotically efficient estimates for the parameters of the Models I and II under Assumption 6.4.2 is the following two-stage Aitken-Zellner estimator, cf. Zellner (1962a).

$$\hat{\underline{\beta}}(s) = \left[\sum_{j=1}^{n} \frac{X_j' X_j}{s_{jj}} \right]^{-1} \left[\sum_{i=1}^{n} \frac{X_i' X_i}{s_{ii}} \underline{b}_i \right] , \tag{6.4.5}$$

where s_{ii} is as shown in (4.3.21).

Two-Stage Aitken-Zellner Estimates

Model I: $$C_{it} = \underset{(0.278)}{7.429} + \underset{(0.001)}{0.906}\, x_{it} \tag{6.4.6}$$

Model II: $$C_{it} = \underset{(0.022)}{0.698}\, C_{i,t-1} + \underset{(0.011)}{0.292}\, x_{it} \; . \tag{6.4.7}$$

We can easily generalize the specification (2) in Assumption 6.4.2, à la Zellner (1962a) and Parks (1967). For the present study this specification seems to be adequate. In order to test whether the specification (1) in Assumption 6.4.2 is consistent with our data, we make use of the statistic (4.3.80) developed by Zellner (1962a). The specification (3) in Assumption 6.4.2 is not adequate to apply this test procedure. We, therefore, change it as follows.

Assumption 6.4.3:

(1) Same as (2) in Assumption 6.4.1.

(2) The explanatory variables C_{it-1} and x_{it} are nonstochastic.

Since we have illustrative purposes in mind, we neglect the fact that the specification (2) in Assumption 6.4.3 is not tenable for the Model II. As regards the distribution of $\underline{u}_i$ we make the following assumption.

Assumption 6.4.4: The $\underline{u}_i$'s are normally distributed.

Under Assumptions 6.4.3 and 6.4.4 and under the null hypothesis

$$H_0: \underline{\beta}_1 = \underline{\beta}_2 = \ldots = \underline{\beta}_n = \underline{\beta} \quad , \tag{6.4.8}$$

the asymptotic distribution of (4.3.80) is F with (n-1)K, (T-K)n d.f.

The value of (4.3.80) is 46.9359 for the Model I and 7.4415 for the Model II. These values are well above the 5 percent value of F with 46 and 144 d.f., so that the data do not support the hypothesis (6.4.8). Consequently, the specification (1) in Assumption 6.4.2 is not valid for our data. The estimates in (6.4.3), (6.4.4), (6.4.6) and (6.4.7) suffer from a specification error of the type discussed in Zellner (1962b). One way of relaxing the condition (6.4.8) is to allow different coefficient vectors for different countries. If these vectors are fixed parameters each country's coefficient vector is to be estimated by using only that country's data. In this case there is a considerable risk of arriving at nonsensical results for some countries if longer time spans are not considered for each country. There is also no way of pooling time series from different countries. If we can assume that the coefficient vectors of different countries are not fixed parameters, but random variables which are distributed with the same mean and the same variance-covariance matrix across countries, we can find an optimal way of pooling data on all these countries. We explore this possibility in the following section.

6.5 Random Coefficient Approach

In this section we estimate the Models I and II under Assumption 4.2.1. If we estimate each equation in (6.2.1) and (6.2.2) separately for each country, we obtain predictions on the $\underline{\beta}_i$'s. The predictions on coefficients and the estimates of their means are given in Table 6.1 along with their standard errors for each country. The standard errors of predictions are obtained by taking the square root of the diagonal elements of $s_{ii}(X_i'X_i)^{-1}$ and the standard errors of estimates are obtained by taking the square root of the diagonal elements of $\hat{\Delta} + s_{ii}(X_i'X_i)^{-1}$ where $\hat{\Delta}$ is as shown in (4.3.24).

All the estimates in Table 6.1 have the right signs. The means of coefficients are not precisely determined. For the present application

$$\begin{array}{c} \\ \hat{\Delta} = \end{array} \begin{array}{cc} \alpha_{0i} & \alpha_{1i} \\ \left[\begin{array}{c} 2864.8 \\ \end{array}\right. & \left.\begin{array}{c} 2.0778 \\ 0.0075 \end{array}\right] \end{array} ; \begin{array}{cc} \gamma_{1i} & \gamma_{2i} \\ \left[\begin{array}{c} 0.025069 \\ \end{array}\right. & \left.\begin{array}{c} -0.020921 \\ 0.017582 \end{array}\right] \end{array} . \qquad (6.5.1)$$

If we assume that $\sigma_{11} = \sigma_{22} = \ldots = \sigma_{nn} = \sigma^2$ and adopt the estimator (4.10.7) we obtain the following estimates.

$$\begin{array}{c} \\ \hat{\Delta}^* = \end{array} \begin{array}{cc} \alpha_{0i} & \alpha_{1i} \\ \left[\begin{array}{c} -5917.3 \\ \end{array}\right. & \left.\begin{array}{c} 72.2 \\ -0.5 \end{array}\right] \end{array} \begin{array}{cc} \gamma_{1i} & \gamma_{2i} \\ \left[\begin{array}{c} -0.409 \\ \end{array}\right. & \left.\begin{array}{c} 0.377 \\ -0.349 \end{array}\right] \end{array} . \qquad (6.5.2)$$

We cannot accept the estimates in (6.5.2) because negative estimates for variances are not plausible. From this we infer that the specification (2) in Assumption 6.4.1 is not valid for our data. The estimator $\hat{\Delta}$ yielded plausible estimates for the variances and there is no reason to reject the specification (2) in Assumption 6.4.2.

The estimates in (6.5.1) indicate a high negative correlation between the coefficients of lagged consumption and income. It has been shown in subsection 4.3(k) that if Assumption 4.3.1 is true, (as $T\to\infty$) the asymptotic distribution of $(n-1) \cdot \underline{\ell}'\hat{\Delta}\,\underline{\ell}/\underline{\ell}'\Delta_0\underline{\ell}$, where $\underline{\ell}$ is a K x 1 vector of arbitrary elements, is χ^2 with (n-1) d.f. Using this statistic we obtained the following 95 percent confidence intervals for the variances of coefficients in Models I and II.

Table 6.1

Predictions on Coefficients in Model I and Model II and Estimates of Their Means for Each Country[30]

	Model I			Model II		
Country	Intercept	Coeff. of Income	Est. Var. of Disturbances (s_{ii})	Coeff. of Lagged Consumption	Coeff. of Income	Est. Var. of Disturbances (s_{ii})
Austria	-23.38 (43.45) [68.94]	0.951 (0.069) [0.111]	154.62	0.992 (0.156) [0.222]	0.048 (0.136) [0.189]	20.987
Australia	258.80 (95.13) [109.15]	0.697 (0.077) [0.115]	285.21	0.393 (0.153) [0.220]	0.557 (0.136) [0.189]	302.29
Belgium	41.79 (52.99) [75.32]	0.848 (0.055) [0.103]	118.40	0.102 (0.204) [0.258]	0.803 (0.177) [0.221]	125.49
Burma	13.05 (15.05) [55.63]	0.655 (0.269) [0.282]	6.53	0.511 (0.268) [0.311]	0.441 (0.235) [0.270]	4.57
Canada	255.28 (80.52) [96.68]	0.740 (0.058) [0.104]	93.58	0.383 (0.146) [0.215]	0.575 (0.132) [0.187]	116.32
Chile	143.08 (82.81) [98.60]	0.630 (0.257) [0.271]	165.17	0.671 (0.460) [0.486]	0.365 (0.486) [0.504]	182.61
China (Taiwan)	45.32 (5.83) [53.84]	0.390 (0.068) [0.110]	1.34	0.857 (0.179) [0.239]	0.143 (0.162) [0.209]	3.08
Columbia	13.89 (40.31) [67.00]	0.900 (0.201) [0.218]	42.63	0.334 (0.237) [0.285]	0.649 (0.227) [0.263]	32.64
Costa Rica	13.04 (24.55) [58.88]	0.902 (0.112) [0.141]	0.781	0.111 (0.066) [0.171]	0.855 (0.063) [0.147]	0.56
Denmark	95.76 (50.32) [73.46]	0.809 (0.045) [0.097]	206.86	0.345 (0.076) [0.181]	0.594 (0.066) [0.148]	74.742
Ecuador	-21.86 (21.25) [57.59]	1.132 (0.154) [0.177]	0.98	0.195 (0.091) [0.182]	0.786 (0.088) [0.159]	0.652

Table 6.1 (Continued)

Country	Model I Intercept	Model I Coeff. of Income	Model I Est. Var. of Disturbances (s_{ii})	Model II Coeff. of Lagged Consumption	Model II Coeff. of Income	Model II Est. Var. of Disturbances (s_{ii})
Finland	94.40 (29.50) [61.12]	0.732 (0.050) [0.099]	103.09	0.338 (0.107) [0.191]	0.599 (0.093) [0.162]	105.2
Honduras	11.60 (32.84) [62.79]	0.863 (0.219) [0.235]	4.29	0.029 (0.157) [0.223]	0.914 (0.147) [0.198]	4.36
Ireland	82.37 (42.64) [68.43]	0.793 (0.066) [0.108]	94.43	0.257 (0.220) [0.271]	0.688 (0.198) [0.238]	124.80
Jamaica	107.92 (38.73) [66.06]	0.603 (0.123) [0.150]	11.02	0.168 (0.156) [0.222]	0.790 (0.145) [0.196]	21.20
Japan	46.50 (9.69) [54.38]	0.658 (0.031) [0.092]	22.15	0.829 (0.100) [0.187]	0.177 (0.076) [0.152]	8.60
Mauritius	93.20 (35.92) [64.46]	0.383 (0.208) [0.225]	28.79	0.686 (0.429) [0.457]	0.295 (0.393) [0.414]	42.86
Netherlands	43.89 (51.82) [73.28]	0.824 (0.061) [0.105]	121.32	0.183 (0.154) [0.221]	0.720 (0.131) [0.186]	109.78
New Zealand	268.36 (134.52) [144.77]	0.702 (0.241) [0.256]	2199.80	0.207 (0.257) [0.301]	0.712 (0.227) [0.263]	2198.20
South Africa	79.86 (77.72) [94.37]	0.730 (0.161) [0.182]	85.85	0.304 (0.203) [0.257]	0.629 (0.178) [0.222]	73.50
Spain	99.26 (33.18) [62.79]	0.640 (0.108) [0.138]	79.34	0.319 (0.160) [0.225]	0.666 (0.149) [0.199]	118.94
Sweden	198.09 (56.72) [77.98]	0.725 (0.046) [0.098]	146.77	0.480 (0.142) [0.213]	0.471 (0.122) [0.180]	153.19

Table 6.1 (Continued)

	Model I			Model II		
Country	Intercept	Coeff. of Income	Est. Var. of Disturbances (s_{ii})	Coeff. of Lagged Consumption	Coeff. of Income	Est. Var. of Disturbances (s_{ii})
Switzerland	327.20 (27.84) [60.33]	0.639 (0.023) [0.089]	58.37	0.740 (0.140) [0.211]	0.249 (0.121) [0.179]	217.44
U.K.	208.68 (45.03) [69.94]	0.754 (0.043) [0.096]	62.26	0.520 (0.216) [0.268]	0.466 (0.202) [0.241]	145.26

[31] The figures given in ( ) are the standard errors of predictions on coefficients and the figures given in [ ] are the standard errors of estimates of the means of coefficients. Due to short span of time series data on each country we could not include a constant term in Model II. When we included a constant term we ran into multi-collinearity problems for some countries.

Model I

Confidence intervals for the variances of coefficients

$$1729.64 < \sigma^2_{\alpha 0} < 5632.44 \ ; \quad 0.0045 < \sigma^2_{\alpha 1} < 0.0147 \tag{6.5.3}$$

where $\sigma^2_{\alpha 0}$ and $\sigma^2_{\alpha 1}$ are the variances of α_{0i} and α_{1i} respectively.

Model II

Confidence intervals for the variances of coefficients

$$0.0151 < \sigma^2_{\gamma 1} < 0.0493 \ ; \quad 0.0106 < \sigma^2_{\gamma 2} < 0.0346 \tag{6.5.4}$$

where $\sigma^2_{\gamma 1}$ and $\sigma^2_{\gamma 2}$ are the variances of γ_{1i} and γ_{2i} respectively. When we adopted the estimator (4.3.25) in our calculations, we obtained the following estimates for the means of coefficients in Models I and II.

Model I

$$\underline{\bar{b}}(\hat{\underline{\theta}}) = \begin{matrix} \hat{\bar{\alpha}}_0 & \hat{\bar{\alpha}}_1 \\ [89.553 & 0.7368]' \\ (10.925) & (0.0168) \end{matrix} \tag{6.5.5}$$

Model II

$$\underline{\bar{b}}(\hat{\underline{\theta}}) = \begin{matrix} \hat{\bar{\gamma}}_1 & \hat{\bar{\gamma}}_2 \\ [0.4696 & 0.4971]' \\ (0.0323) & (0.0270) \end{matrix} \tag{6.5.6}$$

The figures given in the parentheses are the large sample standard errors calculated by taking the square root of the diagonal elements of $\frac{\hat{\Delta}}{n}$.

In order to study the spread of the values of coefficients in Models I and II around their respective mean values, we report the coefficient of variation for each coefficient. The coefficient of variation is measured by $\hat{\sigma}_{\beta_1}/\hat{\bar{\beta}}_1$ where $\hat{\sigma}_{\beta_1}$ is the estimated standard deviation of the first element of $\underline{\beta}_i$ and $\hat{\bar{\beta}}_1$ is the estimated mean value of the first element of $\underline{\beta}_i$.

Coefficients of Models I and II

	α_{0i}	α_{1i}	γ_{1i}	γ_{2i}	
Coefficient of variation:	0.65	0.12	0.34	0.27 .	(6.5.7)

All the coefficients exhibit substantial variation across countries. We cannot possibly assume that the coefficients of Models I and II are the same for all countries.

Now we make a comparative study of estimates in (6.4.3), (6.4.6) and (6.5.5) for Model I and of estimates (6.4.4), (6.4.7) and (6.5.6) for Model II. The analytical formula used to obtain estimates in (6.4.3) or (6.4.4) is given in (6.4.2). The estimates in (6.4.6) or (6.4.7) are based on the formula (6.4.5). The formula (4.3.25) is used to obtain the estimates in (6.5.5) or (6.5.6). These formulae differ in terms of the weighting scheme used to combine the $\underline{b}_i$'s. In the case of (6.4.2), the weight given to $\underline{b}_i$ is $\left[\sum_{j=1}^{n} X_j'X_j\right]^{-1} X_i'X_i$. In the case of (6.4.5) the quantity $\underline{b}_i$ gets a weight equal to $\left[\sum_{j=1}^{n} \frac{X_j'X_j}{s_{jj}}\right]^{-1} \frac{X_i'X_i}{s_{ii}}$. Similarly, the weight given to $\underline{b}_i$ in (4.3.25) is $\left[\sum_{j=1}^{n} \left\{\hat{\Delta} + s_{jj}(X_j'X_j)^{-1}\right\}^{-1}\right]^{-1} \left\{\hat{\Delta} + s_{ii}(X_i'X_i)^{-1}\right\}^{-1}$. The differences in these weights reflect the differences in the assumptions underlying the above formulae. From the results in (6.4.3), (6.4.6), and (6.5.5) [or in (6.4.4), (6.4.7), and (6.5.6)], we can see how the changes in weights in the above manner affect the estimates of parameters in Model I (or in Model II). The weighting scheme in (6.4.2) [or in (6.4.5)] takes into account only the variances and covariances of regressors besides the variances of disturbances. These weights introduce specification errors since the specification (1) in Assumption 6.4.1 is not valid for the situation under study. The particular weighting scheme in (4.3.25) takes into account the variances and covariances of both the coefficients and regressors besides the variances of disturbances. In addition to this, it has another merit. For sufficiently large T each $\underline{b}_i$ in $\bar{\underline{b}}(\hat{\underline{\theta}})$ gets equal weight and $\bar{\underline{b}}(\hat{\underline{\theta}})$ reduces to $\frac{1}{n}\sum_{i=1}^{n} \underline{\beta}_i$ as shown in (4.3.51). In other words, the estimator $\bar{\underline{b}}(\hat{\underline{\theta}})$ is asymptotically equal to the simple arithmetic mean of micro coefficients. As explained in (1.2.25) the simple arithmetic mean of micro coefficients may be viewed as a macro coefficient estimator without an aggregation bias. The weighting scheme used in (4.3.25) yields a macro coefficient estimator without an aggregation bias if T is sufficiently large. Since T is as small as 8 in our study, we cannot draw the conclusion that the estimates in (6.5.5) and (6.5.6) are macro coefficient estimators without an aggregation bias. For both the models the estimates obtained by random coefficient approach are different from those obtained by fixed coefficient approach. We cannot accept the latter because they suffer from a specification error.

In order to test the hypotheses on the parameters of Models I and II, we make use of Assumption 4.3.1. This assumption may not be totally justified here because the coefficients α_{1i}, γ_{1i}, and γ_{2i} take values only between 0 and 1 but not over $-\infty$ to ∞ as a normal variable would take. On the basis of the estimated means and variances of α_{0i}, α_{1i}, γ_{1i} and γ_{2i}, approximate intervals covering 95 percent of the total area under the normal curve are given below.

Model I

$$\Pr\ (-17.49 < \alpha_{0i} < 196.60) \approx 0.95 \tag{6.5.8}$$

$$\Pr\ (\ 0.56 < \alpha_{1i} < \ 0.91) \approx 0.95 \tag{6.5.9}$$

Model II

$$\Pr\ (\ 0.15 < \gamma_{1i} < \ 0.79) \approx 0.95 \tag{6.5.10}$$

$$\Pr\ (\ 0.23 < \gamma_{2i} < \ 0.76) \approx 0.95 \tag{6.5.11}$$

$$\Pr\ (\ 0.91 < \gamma_{1i} + \gamma_{2i} < 1.02) \approx 0.95. \tag{6.5.12}$$

All these intervals in (6.5.9)-(6.5.12) lie within the appropriate ranges of coefficients. If the estimates of the means and variances of coefficients are precisely determined, the probability of values of the coefficients outside the range 0-1 is very small and the normal approximation to the distributions of coefficients may be quite good. Under the null hypothesis that $\underline{\bar{\beta}} = 0$, the value of the statistic (4.3.60) is 484.24 for Model I and 43636.42 for Model II. These values are well above the 5 percent value of F with 2 and 22 d.f. The means of coefficients in Models I and II are significantly different from zero. Even though the individual country regressions did not yield precise results, the pooled estimates in (6.5.5) and (6.5.6) are highly significant.

We have indicated in footnote 28 that we can interpret the results for Model II from the standpoint of permanent income hypothesis. If $e_{it} = \lambda_i e_{it-1} + \epsilon_{it}$ and if ϵ_{it} is independently distributed with mean zero and constant variance, then all the estimates of parameters in Model II, which are given in Table 6.1, are consistent. In a remarkable paper, Zellner and Geisel (1968) showed that this assumption was consistent with the U.S. data which they analyzed. We make a few observations below treating the above assumption as valid for our data. From the relation which connects the coefficients in the Model II with those in the Friedman's model we have

$$\bar{\gamma}_1 = E\lambda_i \text{ and } \bar{\gamma}_2 = E(1 - \lambda_i)\, k_i \quad . \tag{6.5.13}$$

Since we have assumed that γ_{1i} and γ_{2i} are correlated, we cannot possibly assume that λ_i and k_i are uncorrelated. Consequently,

$$\bar{\gamma}_2 = (1 - \bar{\lambda})\, \bar{k} - \text{cov}\,(\lambda_i, k_i) \quad , \tag{6.5.14}$$

where $E\lambda_i = \bar{\lambda}$ and $Ek_i = \bar{k}$. If we have a prior information on cov (λ_i, k_i), we can estimate $\bar{k}$ from (6.5.14) knowing $\bar{\lambda}$. Now $\bar{k} \lesseqgtr \bar{\gamma}_2/(1 - \bar{\lambda})$ according as cov (λ_i, k_i) is negative or positive. We expect the cov (λ_i, k_i) to be negative *a priori* because shorter the horizon (i.e., smaller the value of λ_i) larger the value of k_i. Therefore,

$$\bar{\lambda} = 0.47 \text{ and } \bar{k} < 0.94 \quad . \tag{6.5.15}$$

Notice here that Assumption 4.3.1 implies that Ek_i does not exist. To be consistent with Assumption 4.3.1 we treat $\bar{k}$ as the median of k_i. Further investigation is necessary to ascertain the properties of the median of k_i.

Notice also that the specification (4) in Assumption 4.2.1 reduces to the specification (1) in Assumption 6.4.1 if $\Delta = 0$. Now we test the hypothesis,

$$H_0: \quad \Delta = 0, \text{ given that } E\underline{\beta}_i = \bar{\underline{\beta}} \quad . \tag{6.5.16}$$

The value of the statistic (4.3.70) is 104.15 for Model I and 34.47 for Model II. These values are also well above the 5 percent value of χ^2 with 3 d.f. Therefore, we cannot accept the hypothesis (6.5.16). This result clearly indicates that the fixed coefficient approach presented in Section 6.4 is not appropriate to the analysis of cross-country data on aggregate consumer expenditures. If the assumption that $E\underline{\beta}_i = \overline{\underline{\beta}}$ is false, the data on all these countries cannot be pooled. The specification errors due to the falsity of the assumption $E\underline{\beta}_i = \overline{\underline{\beta}}$ is unavoidable in cross-section analysis.

6.6 Conclusions

Taking comparable data on per capita real consumption expenditures and per capita real income for each of 24 countries over a period of 8 years from 1956-63, we estimated Kaynesian and habit persistence (or permanent income) type consumption functions. Allowing the coefficient vector of each model to vary randomly across countries but assuming that it is distributed with the same mean and the same variance-covariance matrix, we estimated the above two models. This procedure gave us plausible results. We have shown that results from the traditional fixed coefficient approach suffer from a common specification error discussed in Zellner (1962b). When the units in a cross-section are not homogeneous with regard to a coefficient vector, the data on these units can be pooled to estimate a single relation if the coefficient vector is random across units and distributed with the same mean. Our study suffers from a limitation imposed by the short span of our time series sample. The critical assumption which permits pooling of time series from different countries under the random coefficient approach is that the means of coefficient vectors are the same for all countries. If this assumption is wrong, the results (6.5.5) and (6.5.6) suffer from a specification error.

CHAPTER VII

MISCELLANEOUS TOPICS

7.1 Introduction

We discuss in this chapter some problems related to RCR models. These problems concern identification, the use of prior information, and the presence of random and nonrandom coefficients in the relation. We take up these problems one by one and discuss them from the standpoint of RCR models. Since the RCR models have a special appeal in the analysis of panel data, it is important to establish the properties of estimators derived in Chapter IV as completely as possible. The present chapter represents an effort in this direction.

In Section 7.2 we discuss the identification problem; in Section 7.3 we discuss the use of prior information, and in Section 7.4 we present a short summary of the study in this chapter.

7.2 Identification

The identification problem arising in fixed coefficient models has been discussed in various terminologies and formulations by quantitative thinkers in several fields. A comprehensive survey of many of these formulations of the identification problem has been given in Fisher (1966). Similar problems can also arise in RCR models. The problems of identification in RCR models have some similarities with those in Bayesian analysis, cf. Drèze (1962), and Zellner (1965).

The model we consider here is the one described in (4.2.1) for the ith individual. We utilize Assumption 4.2.1.

We review briefly below some of the familiar concepts of identification with reference to (4.2.1). The model (4.2.1) defines a group of structures, $\Phi_s = \left\{\underline{\bar{\beta}}, \underline{\theta}\right\}$. A given structure generates one and only one joint probability density function (pdf), $p(\underline{y})$, of the observed variable y_{ti}. On the other hand, there may be several structures Φ_s generating the same distribution $p(\underline{y})$. If two or more structures generate the same probability distribution of the observed variable, the structures are said to be observationally equivalent. If a parameter has the same value in all equivalent structures, it is said to be identifiable. In other words, a parameter is identifiable if it can be uniquely determined from a knowledge of the probability distribution of the observed variable. If a parameter is not identifiable, no consistent estimate of the parameter will exist.

We can easily apply the concept of identification to our model (4.2.1) in terms of the marginal pdf of observations on the dependent variable. Suppose that we have two models, say M_I with its parameter vector $(\underline{\bar{\beta}}', \underline{\theta}')'_I$ and joint density function $f(\underline{y}|M_I)$ and M_{II} with its parameter vector $(\underline{\bar{\beta}}', \underline{\theta}')'_{II}$ and joint density function $g(\underline{y}|M_{II})$. Then the marginal distribution of $\underline{y}$, given model M_I, is

$$f(\underline{y}|M_I) = \int_{R_\delta} f(\underline{u},\underline{\delta}|M_I)|J|d\underline{\delta} \quad , \tag{7.2.1}$$

where $\underline{y}$ is an observation vector on the dependent variable, $|J|$ is the Jacobian of the transformation from $\underline{u},\underline{\delta}$ to $\underline{y},\underline{\delta}$, and R_δ is the range of $\underline{\delta}$ defined by M_I.

Similarly, for model M_{II} with its associated parameter vector $(\underline{\bar{\beta}}', \underline{\theta}')'_{II}$,

$$g(\underline{y}|M_{II}) = \int_{R_\delta} g(\underline{u},\underline{\delta}|M_{II})|J|d\underline{\delta} \quad . \tag{7.2.2}$$

Then M_I and M_{II} are defined to be observationally equivalent if and only if

$$f(\underline{y}|M_I) = g(\underline{y}|M_{II}) \quad . \tag{7.2.3}$$

When (7.2.3) holds for M_I and M_{II} we cannot decide whether M_I or M_{II} explains the data. If $f(\underline{u},\underline{\delta}|M_I)$ is not invariant to the class of transformations considered, then M_I and M_{II} are not equivalent.

The above discussion on equivalent structures is in fact related to the familiar Fisherian concept of information defined in Rao (1965a, p. 263). By information on an unknown parameter $\underline{\theta}$ contained in a random variable, we mean the extent to which uncertainty regarding the unknown value of $\underline{\theta}$ is reduced as a consequence of an observed value of the random variable. If the variable takes a unique value with probability one corresponding to each value of $\underline{\theta}$, we have a situation where the random variable has the maximum information. On the other hand, if the random variable has the same distribution for all values of $\underline{\theta}$, there is no way of making statements about the value of $\underline{\theta}$ on the basis of an observed value of such a random variable. There is no information (about $\underline{\theta}$) contained in such a random variable. This is the situation described by (7.2.3). The sensitiveness of a random variable with respect to a parameter may then be judged by the extent to which its distribution is altered by a change in the value of the parameter. If a random variable is highly sensitive to infinitesimal changes in the value of $\underline{\theta}$, the observed value of the random variable contains some information on $\underline{\theta}$. In other words, if $\underline{y}$ has different distributions for different values of $\underline{\theta}$, an observation on $\underline{y}$ contains some information on $\underline{\theta}$. If $f(\underline{y};\bar{\underline{\beta}}_I,\underline{\theta}_I)$ and $g(\underline{y};\bar{\underline{\beta}}_{II},\underline{\theta}_{II})$ denote different probability densities of $\underline{y}$ for two different values of the parameter $(\bar{\underline{\beta}}',\underline{\theta}')'$, the difference in the distributions could be measured by some distance functions, cf. Rao (1965a, p. 271). Zellner (1965) has indicated that Kullback's measure of discrimination information can be used to study whether a given body of data explains one model or the other. If there are significant differences between the densities of $\underline{y}$ for different values of $(\underline{\theta}',\bar{\underline{\beta}})'$, the parameters are identified.

Since the coefficients are random in (4.2.1) we cannot impose the exact linear restrictions on coefficients in the same way as we do in fixed coefficient models to achieve identification. We have to properly redefine the concept of identification to suit our model. We can say, in the case of RCR models, that certain values of coefficients are a priori less probable than others, although not ruled out completely, cf. Drèze (1962). If we know a priori that it is less probable for any coefficient in a RCR model to take values other than zero, we can impose the restrictions that the coefficient has zero mean and zero variance. Let us call these as double zero restrictions. They serve the same purpose as zero restrictions in fixed coefficient models. If we know a priori the range of the distribution of some or all the elements of $\underline{\beta}_i$, we can use this information in achieving identification in some cases.

7.3 Incorporation of Prior Information in the Estimation of RCR Models

In Section 4.3 of Chapter IV we have discussed the procedures of estimating the parameters of the model (4.2.1) under Assumption 4.2.1. We have shown that the estimator $\bar{\underline{b}}(\hat{\underline{\theta}})$ in (4.3.25) is a consistent and asymptotically efficient estimator of $\bar{\underline{\beta}}$, the mean of the random coefficient vector $\underline{\beta}_i$. The estimator $\bar{\underline{b}}(\hat{\underline{\theta}})$ is based on the information supplied by panel data. However, there are many occasions in which the researcher possesses, in addition to the sample, a priori information about the parameters of the model. In the case of (4.2.1) the unknown parameters are $\bar{\underline{\beta}}$, and $\underline{\theta}$ defined in (4.3.3). We consider in this section prior information on $\bar{\underline{\beta}}$, which may be of stochastic or nonstochastic nature. For example, nonstochastic prior information in the case of a demand function for an economic good derived from the theory of utility maximization subject to a budget constraint is that the demand function is homogeneous of degree zero in prices and income. If we are dealing with a consumption function, stochastic prior information is that the marginal propensity to consume, being a random variable in RCR models, takes values between zero and one. In the following paragraphs we combine prior information with sample information in estimating $\bar{\underline{\beta}}$ and show that the use of nonstochastic prior information improves the efficiency of an estimator of $\bar{\underline{\beta}}$. The use of stochastic prior information is likely to improve the efficiency of an estimator of $\bar{\underline{\beta}}$ in small samples, cf. Swamy and Mehta (1969). We first consider the use of nonstochastic prior information.

(a) Use of Nonstochastic Prior Information

In (4.3.15) we have shown that by minimizing the quadratic form

$$\sum_{i=1}^{n} (\underline{b}_i - \bar{\underline{\beta}})'[\Delta + \sigma_{ii}(X_i'X_i)^{-1}]^{-1} (\underline{b}_i - \bar{\underline{\beta}}) \quad , \qquad (7.3.1)$$

we obtain the estimator $\bar{\underline{b}}(\underline{\theta})$, which is the BLUE of $\bar{\underline{\beta}}$. In this procedure we have not imposed any constraint on $\bar{\underline{\beta}}$.

Suppose that we have extraneous information in the form of exact linear restrictions on the means of coefficients;

$$\underline{r} = R\underline{\bar{\beta}} \tag{7.3.2}$$

where $\underline{r}$ is a Gx1 known vector and R is a GxK known matrix of rank $G \le K$, so that we have G independent restrictions on the elements of $\underline{\bar{\beta}}$. Examples of exact linear restrictions on coefficients are given in Goldberger (1964, p. 256).

To incorporate this information we suggest the following procedure:

$$\min_{\underline{\bar{\beta}}} \sum_{i=1}^{n} (\underline{b}_i - \underline{\bar{\beta}})'[\Delta + \sigma_{ii}(X_i'X_i)^{-1}]^{-1} (\underline{b}_i - \underline{\bar{\beta}})$$

subject to

$$R\underline{\bar{\beta}} = \underline{r} \quad .$$

Therefore we minimize, following Goldberger (1964, p. 256),

$$Q_r = \sum_{i=1}^{n} (\underline{b}_i - \underline{\bar{\beta}})'[\Delta + \sigma_{ii}(X_i'X_i)^{-1}]^{-1} (\underline{b}_i - \underline{\bar{\beta}}) - 2\underline{\lambda}'(R\underline{\bar{\beta}}-\underline{r}) \tag{7.3.3}$$

with respect to $\underline{\bar{\beta}}$ and $\underline{\lambda}$, where $\underline{\lambda}$ is a Gx1 vector of Lagrangian multipliers. Setting the derivative of Q_r with respect to $\underline{\bar{\beta}}$ equal to zero gives

$$\frac{1}{2}\frac{\partial Q_r}{\partial \underline{\bar{\beta}}} = - \sum_{i=1}^{n} (\Delta + \sigma_{ii}(X_i'X_i)^{-1}]^{-1} (\underline{b}_i - \underline{\bar{\beta}}) - R'\underline{\lambda} = 0 \quad . \tag{7.3.4}$$

Hence,

$$\underline{\bar{b}}_r(\underline{\theta}) = \underline{\bar{b}}(\underline{\theta}) + C(\underline{\theta})R'\underline{\lambda} \quad , \tag{7.3.5}$$

where $C(\underline{\theta})$ is as defined in (4.3.8) and $\underline{\bar{b}}(\underline{\theta})$ shown in (4.3.6) is the unconstrained estimator of $\underline{\bar{\beta}}$.

Premultiplying (7.3.5) by R and imposing the restriction $R\underline{\bar{b}}_r(\underline{\theta}) = \underline{r}$ gives

$$\underline{r} = R\underline{\bar{b}}_r(\underline{\theta}) = R\underline{\bar{b}}(\underline{\theta}) + RC(\underline{\theta})R'\underline{\lambda} \quad . \tag{7.3.6}$$

Hence,

$$\hat{\underline{\lambda}} = [RC(\underline{\theta})R']^{-1} [\underline{r} - R\underline{\bar{b}}(\underline{\theta})] \quad . \tag{7.3.7}$$

Inserting this back into (7.3.5) gives

$$\underline{\bar{b}}_r(\underline{\theta}) = \underline{\bar{b}}(\underline{\theta}) + C(\underline{\theta})R'[RC(\underline{\theta})R']^{-1}[\underline{r} - R\underline{\bar{b}}(\underline{\theta})] \quad . \qquad (7.3.8)$$

It is seen that the constrained estimator $\underline{\bar{b}}_r(\underline{\theta})$ differs from the unconstrained estimator $\underline{\bar{b}}(\underline{\theta})$ by a linear function of the amount $[\underline{r} - R\underline{\bar{b}}(\underline{\theta})]$ by which $\underline{\bar{b}}(\underline{\theta})$ fails to satisfy the restrictions.

The sampling properties of $\underline{\bar{b}}_r(\underline{\theta})$ may be gotten as follows:

$$\underline{\bar{b}}_r(\underline{\theta}) - \underline{\bar{\beta}} = \underline{\bar{b}}(\underline{\theta}) - \underline{\bar{\beta}} + C(\underline{\theta})R'[RC(\underline{\theta})R']^{-1}[\underline{r} - R\underline{\bar{\beta}} + R\underline{\bar{\beta}} - R\underline{\bar{b}}(\underline{\theta})]. \qquad (7.3.9)$$

Using the constraint $\underline{r} - R\underline{\bar{\beta}} = 0$ we obtain

$$\underline{\bar{b}}_r(\underline{\theta}) - \underline{\bar{\beta}} = \underline{\bar{b}}(\underline{\theta}) - \underline{\bar{\beta}} - C(\underline{\theta})R'[RC(\underline{\theta})R']^{-1} R [\underline{\bar{b}}(\underline{\theta}) - \underline{\bar{\beta}}] \quad . \qquad (7.3.10)$$

We take expectations on both sides of (7.3.10) to obtain

$$E\underline{\bar{b}}_r(\underline{\theta}) = \underline{\bar{\beta}} \quad . \qquad (7.3.11)$$

We can easily show that the variance-covariance matrix of $\underline{\bar{b}}_r(\underline{\theta})$ is

$$V [\underline{\bar{b}}_r(\underline{\theta})] = V [\underline{\bar{b}}(\underline{\theta})] - C(\underline{\theta})R'[RC(\underline{\theta})R']^{-1} RC(\underline{\theta}) \quad . \qquad (7.3.12)$$

By comparing (7.3.12) with (4.3.8) we can easily see that the use of prior information improves the efficiency of an estimator of $\underline{\bar{\beta}}$. Since $\underline{\theta}$ is usually unknown, (7.3.8) cannot be adopted in practice. We, therefore, suggest the following estimator based on the estimator $\underline{\hat{\theta}}$ of $\underline{\theta}$.

$$\underline{\bar{b}}_r(\underline{\hat{\theta}}) = \underline{\bar{b}}(\underline{\hat{\theta}}) + C(\underline{\hat{\theta}})R'[RC(\underline{\hat{\theta}})R']^{-1}[\underline{r} - R\underline{\bar{b}}(\underline{\hat{\theta}})] \quad . \qquad (7.3.13)$$

On the basis of the results in subsection 4.3(g) of Chapter IV we have

$$\text{plim} \left|\underline{\bar{b}}_r(\underline{\hat{\theta}}) - \underline{m}_\beta + \frac{S_\beta}{n-1} R'(R \frac{S_\beta}{n-1} R')^{-1} R(\underline{m}_\beta - \underline{\bar{\beta}})\right| = 0 \quad . \qquad (7.3.14)$$

According to the limit theorem in Rao (1965a, p. 101) the asymptotic distribution of $(\underline{\bar{b}}_r(\underline{\hat{\theta}}) - \underline{\bar{\beta}})$ as $T\to\infty$ and n is fixed, is the same as that of the variable

$$\left[I - \frac{S_\beta}{n-1} R' \left(R \frac{S_\beta}{n-1} R'\right)^{-1} R\right] (\underline{m}_\beta - \underline{\bar{\beta}}) \quad . \qquad (7.3.15)$$

Fortunately, it is not difficult to derive the finite sample moments of (7.3.15) under Assumption 4.3.1.

It is shown in the Appendix to this chapter that the random vector (7.3.15) has mean equal to zero and variance-covariance matrix equal to

$$\left[\frac{\Delta}{n} - \frac{\Delta}{n} R'(R\Delta R')^{-1} R\Delta\right] \frac{n-2}{n-G-2} \quad . \tag{7.3.16}$$

When n is sufficiently large such that $n-2 \approx n-G-2$, (7.3.16) is approximately equal to

$$\frac{\Delta}{n} - \frac{\Delta}{n} R'(R\Delta R')^{-1} R\Delta \quad . \tag{7.3.17}$$

It is easy to see that there has been a gain in asymptotic efficiency as a result of the use of prior information. First, $\frac{\Delta}{n}$ which is the asymptotic variance-covariance matrix of the unrestricted estimator $\underline{\bar{b}}(\hat{\underline{\theta}})$ is taken to be positive definite. Then since R' is a KxG matrix of rank G, it follows from a result given in Goldberger (1964, p. 35) that $R\Delta R'$ is positive definite, and then it follows that $(R\Delta R')^{-1}$ is positive definite. Further, the second matrix on the r.h.s. of (7.3.17), namely $(R\Delta)'(R\Delta R')^{-1}R\Delta$ is nonnegative definite according to a result in Goldberger (1964, p. 37). Thus, the asymptotic variance-covariance matrix of $\underline{\bar{b}}_r(\hat{\underline{\theta}})$ is equal to $\frac{\Delta}{n}$ less a nonnegative definite matrix; that is, each diagonal element of the asymptotic variance-covariance matrix of $\underline{\bar{b}}_r(\hat{\underline{\theta}})$ is less than or equal to the corresponding element of $\frac{\Delta}{n}$. We conclude that the asymptotic variance of each element of $\underline{\bar{b}}_r(\hat{\underline{\theta}})$ is less than or equal to the asymptotic variance of the corresponding element of $\underline{\bar{b}}(\hat{\underline{\theta}})$. For finite n, the asymptotic variances of the elements of $\underline{\bar{b}}_r(\hat{\underline{\theta}})$ are given by the diagonal elements of (7.3.16) which are larger than the corresponding diagonal elements of (7.3.17).

(b) Use of Stochastic Prior Information on the Means of Coefficients

Now suppose that the extraneous information on $\underline{\bar{\beta}}$ is not exact but rather consists of unbiased estimates of some or all the elements of $\underline{\bar{\beta}}$ obtained from introspection. We now turn to a method that uses such an extraneous information together with the sample information to estimate $\underline{\bar{\beta}}$ efficiently. Once again let us start with the model (4.2.1). We utilize Assumption 4.2.1. Suppose we have $\underline{r}$ such that it is an unbiased estimate of $\underline{\bar{\beta}}_1$ which is a subvector of $\underline{\bar{\beta}}$. That is,

$$\underset{(G\times 1)}{\underline{r}} = \underset{(G\times 1)}{\underline{\bar{\beta}}_1} + \underset{(G\times 1)}{\underline{v}} \qquad (G \le K) \quad , \tag{7.3.18}$$

where $E\underline{v} = 0$, $E\underline{u}_i\underline{v}' = 0$ for all i, and $E\underline{v}\underline{v}' = \psi$ which is assumed to be known a priori and is nonsingular.

We can rewrite (7.3.18) as

$$\underset{(G\times 1)}{\underline{r}} = \underset{(G\times K)}{R} \; \underset{(K\times 1)}{\underline{\bar{\beta}}} + \underset{(G\times 1)}{\underline{v}} \quad , \tag{7.3.19}$$

where R is known, cf. Theil and Goldberger (1961) and Goldberger (1964, p. 261).

We write the extraneous and sample information together as

$$\begin{pmatrix} \underline{y} \\ \underline{r} \end{pmatrix} = \begin{pmatrix} X \\ R \end{pmatrix} \underline{\bar{\beta}} + \begin{pmatrix} D(X)\underline{\delta}+\underline{u} \\ \underline{v} \end{pmatrix} \quad . \tag{7.3.20}$$

The symbols $\underline{v}$, R, $\underline{r}$ are explained in (7.3.19) and the remaining symbols are explained in (4.3.2). We have for the variance-covariance matrix of the extended random vector

$$E\begin{pmatrix} D(X)\underline{\delta}+\underline{u} \\ \underline{v} \end{pmatrix}[(D(X)\underline{\delta}+\underline{u})', \underline{v}'] = \begin{pmatrix} H(\underline{\theta}) & 0 \\ 0 & \psi \end{pmatrix} \quad , \tag{7.3.21}$$

where $H(\underline{\theta})$ is as shown in (4.3.3).

Applying Aitken's generalized least squares to (7.3.20) we get the BLUE of $\underline{\bar{\beta}}$ as

$$\underline{\bar{b}}_c(\underline{\theta}) = [X'H(\underline{\theta})^{-1}X + R'\psi^{-1}R]^{-1}(X'H(\underline{\theta})^{-1}\underline{y} + R'\psi^{-1}\underline{r}) \tag{7.3.22}$$

$$= \left[\left(\sum_{j=1}^{n} X_j' \left\{X_j \Delta X_j' + \sigma_{jj} I\right\}^{-1} X_j\right) + R'\psi^{-1}R\right]^{-1}$$

$$\cdot \left[\left(\sum_{i=1}^{n} X_i' \left\{X_i \Delta X_i' + \sigma_{ii} I\right\}^{-1} \underline{y}_i\right) + R'\psi^{-1}\underline{r}\right] \quad . \tag{7.3.23}$$

Making use of (4.3.5) we have

$$\bar{\underline{b}}_c(\underline{\theta}) = \left[\sum_{j=1}^{n} \left\{\Delta + \sigma_{jj}(X_j'X_j)^{-1}\right\}^{-1} + R'\psi^{-1}R\right]^{-1}$$

$$\cdot \left[\sum_{i=1}^{n} \left\{\Delta + \sigma_{ii}(X_i'X_i)^{-1}\right\}^{-1} \underline{b}_i + R'\psi^{-1}\underline{r}\right] \quad . \qquad (7.3.24)$$

We can write

$$\bar{\underline{b}}_c(\underline{\theta}) = \bar{\underline{\beta}} + \left[\sum_{j=1}^{n} \left\{\Delta + \sigma_{jj}(X_j'X_j)^{-1}\right\}^{-1} + R'\psi^{-1}R\right]^{-1}$$

$$\cdot\left[\sum_{i=1}^{n}\left\{\Delta + \sigma_{ii}(X_i'X_i)^{-1}\right\}^{-1}[\underline{\delta}_i + (X_i'X_i)^{-1}X_i'\underline{u}_i] + R'\psi^{-1}\underline{v}\right] \quad .$$

$$(7.3.25)$$

Taking expectations on both sides of (7.3.25) we have

$$E\bar{\underline{b}}_c(\underline{\theta}) = \bar{\underline{\beta}} \quad . \qquad (7.3.26)$$

We can easily verify that the variance-covariance matrix of $\bar{\underline{b}}_c(\underline{\theta})$ is

$$V[\bar{\underline{b}}_c(\underline{\theta})] = \left[\sum_{i=1}^{n} \left\{\Delta + \sigma_{ii}(X_i'X_i)^{-1}\right\}^{-1} + R'\psi^{-1}R\right]^{-1} \quad . \qquad (7.3.27)$$

We have shown the variance of $\bar{\underline{b}}(\underline{\theta})$, which is an unconstrained estimator, in (4.3.8). Since $R'\psi^{-1}R$ is a nonnegative definite matrix, each diagonal element of (7.3.27) is less than or equal to the corresponding diagonal element of (4.3.8). Thus, the use of stochastic prior information improves the efficiency of an estimator of $\bar{\underline{\beta}}$.

Despite all these optimal properties of $\bar{\underline{b}}_c(\underline{\theta})$, it cannot be adopted in practice because Δ and σ_{ii} are usually unknown. Following the usual practice we replace Δ and σ_{ii} in $\bar{\underline{b}}_c(\underline{\theta})$ by their respective unbiased estimators and study the consequences of such a procedure. The unbiased estimators of σ_{ii} and Δ are given by (4.3.21) and (4.3.24), respectively. Now we will consider the following estimator.

$$\bar{\underline{b}}_c(\hat{\underline{\theta}}) = \left[\sum_{j=1}^{n} \left\{\hat{\Delta} + s_{jj}(X_j'X_j)^{-1}\right\}^{-1} + R\,\psi^{-1}R\right]^{-1}$$

$$\cdot\left[\sum_{i=1}^{n} \left\{\hat{\Delta} + s_{ii}(X_i'X_i)^{-1}\right\}^{-1} \underline{b}_i + R'\psi^{-1}\,\underline{r}\right] \quad . \qquad (7.3.28)$$

Using again the results in subsection 4.3(g) of Chapter IV, we can easily show that the estimators $\underline{\bar{b}}_c(\underline{\hat{\theta}})$ and $\underline{\bar{b}}(\underline{\hat{\theta}})$ will have the same asymptotic variance-covariance matrix if the stochastic prior information is dominated by sample information in large samples. In small samples the estimator $\underline{\bar{b}}_c(\underline{\hat{\theta}})$ is likely to be more efficient than $\underline{\bar{b}}_c(\underline{\hat{\theta}})$, cf. Swamy and Mehta (1969).

7.4 Conclusions

In this chapter we discussed the concept of identification within the framework of RCR models. Zero restrictions on both the means and variances of coefficients, and restrictions on the ranges of the distributions of coefficients help us in identifying RCR models.

By incorporating nonstochastic prior information in the estimation of parameters of RCR models we can improve the asymptotic efficiency of an estimator of the mean of a random coefficient vector. The use of stochastic prior information may increase the efficiency of an estimator of the mean vector in small samples.

APPENDIX TO CHAPTER VII

7.A. Derivation of the Mean and the Variance-Covariance Matrix of the Limiting Distribution of Constrained Estimator

Consider the random variable (7.3.15). Under Assumption 4.3.1 the sample mean $\underline{m}_\beta$ is independent of the sample variance-covariance matrix S_β. The expectation of (7.3.15) conditional on S_β is $\underline{\bar{\beta}}$. Therefore the mean of the limiting distribution of $\underline{\bar{b}}_r(\underline{\theta})$ as $T\to\infty$ and n is fixed is $\underline{\bar{\beta}}$. The variance-covariance matrix of (7.3.15) conditional on S_β is

$$\left\{I - (n-1)^{-1}S_\beta R'[R(n-1)^{-1}S_\beta R']^{-1}R\right\} \frac{\Delta}{n} \left\{I - (n-1)^{-1}S_\beta R' \cdot[R(n-1)^{-1}S_\beta R']^{-1}R\right\}' \quad . \tag{7.A.1}$$

The unconditional variance-covariance matrix of (7.3.15) is simply the expectation of (7.A.1) over the distribution of S_β. We will first express (7.A.1) in a form from which it will be easy to evaluate the variance-covariance matrix of (7.A.1).

We note that since Δ is assumed to be positive definite, we can express

$$\Delta = CC' \quad , \tag{7.A.2}$$

where C is a lower triangular matrix with nonnegative diagonal elements, cf. Rao [1965a, p. 55 (3.6)].

We can also find an orthogonal matrix, O_G say, of order G such that

$$O_G' R\Delta R' O_G = D^2 \quad , \tag{7.A.3}$$

where D^2 is a diagonal matrix. Let

$$C_5 = D^{-1}O_G' RC \quad . \tag{7.A.4}$$

We can easily verify that $C_5C_5' = I_G$. To the row-orthonormal matrix C_5 we can add K-G orthonormal rows C_6 such that $C_6C_6' = I_{K-G}$, and relate C_5 and C_6 through $C_5'C_5 = I_K - C_6'C_6$ by letting $[C_5', C_6']$ be an orthogonal matrix. Notice that C_6 is null if K=G. Now express

$$S^*_\beta = C^{-1} S_\beta C'^{-1} \quad , \tag{7.A.5}$$

such that S^*_β is Wishart distributed with n-1 d.f. and $E(n-1)S^*_\beta = I$, cf. Rao (1965a, p. 508, 11.6). It follows from (7.A.4) that $RC = O_G D C_5$. With these results we can write (7.A.1) as

$$\begin{aligned}
&\frac{\Delta}{n} - \frac{C}{n} C_5' D' O_G' (O_G D C_5 S^*_\beta C_5' D' O_G')^{-1} O_G D C_5 S^*_\beta C' \\
&- C S^*_\beta C_5' D' O_G' (O_G D C_5 S^*_\beta C_5' D' O_G')^{-1} O_G D C_5 \frac{C'}{n} \\
&+ C S^*_\beta C_5' D' O_G' (O_G D C_5 S^*_\beta C_5' D' O_G')^{-1} O_G \frac{D^2}{n} O_G \\
&\cdot (O_G D C_5 S^*_\beta C_5' D' O_G')^{-1} O_G D C_5 S^*_\beta C' \quad .
\end{aligned} \tag{7.A.6}$$

This simplifies to

$$\begin{aligned}
&\frac{\Delta}{n} - \frac{C}{n} C_5' (C_5 S^*_\beta C_5')^{-1} C_5 S^*_\beta C' - C S^*_\beta C_5' (C_5 S^*_\beta C_5')^{-1} C_5 \frac{C'}{n} \\
&+ \frac{C}{n} S^*_\beta C_5' (C_5 S^*_\beta C_5')^{-1} (C_5 S^*_\beta C_5')^{-1} C_5 S^*_\beta C' \quad .
\end{aligned} \tag{7.A.7}$$

Now replace $S^*_\beta C'$ by $S^*_\beta (C_5' C_5 + C_6' C_6) C'$ and cancel some terms to obtain

$$\frac{C C_6' C_6 C'}{n} + \frac{C}{n} C_6' C_6 S^*_\beta C_5' (C_5 S^*_\beta C_5')^{-1} (C_5 S^*_\beta C_5')^{-1} C_5 S^*_\beta C_6' C_6 C' \quad . \tag{7.A.8}$$

The matrices $(C_5 S^*_\beta C_5')^{-1}$ and $C_5 S^*_\beta C_6'$ in (7.A.8) have very simple distributional forms which can be derived from that of the Wishart matrix

$$\begin{pmatrix} C_5 \\ C_6 \end{pmatrix} S^* \, (C_5', C_6') \quad , \tag{7.A.9}$$

and its triangular factorization

$$HH' = \begin{bmatrix} H_{11} & 0 \\ H_{21} & H_{22} \end{bmatrix} \begin{bmatrix} H_{11}' & H_{21}' \\ 0 & H_{22}' \end{bmatrix} \tag{7.A.10}$$

where H_{11} is a GxG matrix, H_{22} is a (K-G) x (K-G) matrix, H_{21} is a (K-G) x G matrix. Equating the matrices in (7.A.9) and (7.A.10), we can write

$$C_6 S^*_\beta C_5' (C_5 S^*_\beta C_5')^{-1} (C_5 S^*_\beta C_5')^{-1} C_5 S^*_\beta C_6' = H_{21} H_{11}^{-1} H_{11}'^{-1} H_{21}' \quad . \tag{7.A.11}$$

The elements of H are independent. The elements of H_{21} are distributed as $N_1(0,1)$. The square of the ith diagonal element of H_{11} is distributed as χ^2 with n-i d.f., cf. Rao (1965a, p. 507, Problem 11.3). Consequently,

$$EH_{21}H_{11}^{-1}H_{11}'^{-1}H_{21}' = G\,[\mathrm{tr}\; E(H_{11}'H_{11})^{-1}]\, I_{K-G} \quad . \tag{7.A.12}$$

Each diagonal element of $(H_{11}'H_{11})^{-1}$ is the inverse of a χ^2 variable with n-G d.f. and has expectation equal to $(n-G-2)^{-1}$. Consequently, the expectation of (7.A.8) is

$$\frac{CC_6'C_6C'}{n}\left(1 + \frac{G}{n-G-2}\right) = \left[\frac{\Delta}{n} - \frac{\Delta}{n} R'(R\Delta R')^{-1}R\Delta\right]\frac{n-2}{n-G-2} \quad .$$

The above proof is heavily based on the studies by Williams (1967) and Rao (1967).

BIBLIOGRAPHY

Aitken, A. C. (1934-35): "On Least Squares and Linear Combination of Observations", Proceedings of the Royal Society of Edinburgh, 55, pp. 42-8.

Anderson, T. W. (1958): An Introduction to Multivariate Statistical Analysis, New York: John Wiley & Sons.

Balestra, P. and M. Nerlove (1966): "Pooling Cross-Section and Time Series Data in the Estimation of a Dynamic Model: The Demand for Natural Gas", Econometrica, 34, pp. 585-612.

Bement, T. R. and J. S. Williams (1969): "Variance of Weighted Regression Estimators When Sampling Errors are Independent and Heteroskedastic", Journal of the American Statistical Association, 64, pp. 1369-82.

Boot, J.C.G. and G. M. de Wit (1960): "Investment Demand: An Empirical Contribution to the Aggregation Problem", International Economic Review, 1, pp. 3-30.

Box, G.E.P. (1954): "Some Theorems on Quadratic Forms Applied in the Study of Analysis of Variance Problems, I. Effect of Inequality of Variance in the One-Way Classification", Annals of Mathematical Statistics, 25, pp. 290-302.

Brown, T. M. (1952): "Habit Persistence and Lags in Consumer Behavior", Econometrica, 20, pp. 355-71.

Craig, C. C. (1936): "On the Frequency Function of xy", Annals of Mathematical Statistics, 7, pp. 1-15.

__________ (1942): "On the Frequency Distributions of the Quotient and of the Product of Two Statistical Variables", American Mathematical Monthly, 49, pp. 24-32.

Cramér, H. (1946): Mathematical Methods of Statistics, Princeton: Princeton University Press.

Drèze, J. (1962): "The Bayesian Approach to Simultaneous Equation Estimation", O.N.R. Research Memorandum No. 67, The Technical Institute, Northwestern University, Evanston, Illinois.

Dwyer, P. S. (1967): "Some Applications of Matrix Derivatives in Multivariate Analysis", Journal of the American Statistical Association, 62, pp. 607-25.

Fieller, E. C. (1932): "The Distribution of the Index in a Normal Bivariate Population", Biometrika, 24, pp. 228-40.

Fisher, F. M. (1966): The Identification Problem in Econometrics, New York: McGraw-Hill Book Company.

Fisk, P. R. (1967): "Models of the Second Kind in Regression Analysis", Journal of the Royal Statistical Society Series B, 28, pp. 266-81.

Fraser, D.A.S. and I. Guttman (1956): "Tolerance Regions", Annals of Mathematical Statistics, 26, pp. 162-79.

Friedman, M. (1957): A Theory of the Consumption Function, Princeton: Princeton University Press.

Goldberger, A. S. (1964): Econometric Theory, New York: John Wiley & Sons.

Goodman, L. A. (1960): "On the Exact Variance of Products", Journal of the American Statistical Association, 55, pp. 708-13.

Gradshteyn, I. S. and I. M. Ryzhik (1965): Tables of Integrals, Series, and Products, New York: Academic Press.

Gould, J. P. (1967): "Market Value and the Theory of Investment of the Firm", American Economic Review, 57, pp. 910-13.

Graybill, F. A. (1961): An Introduction to Linear Statistical Models, Vol. I, New York: McGraw-Hill Book Company.

Griliches, Z. and N. Wallace (1965): "The Determinants of Investment Revisited", International Economic Review, 6, pp. 311-29.

Grunfeld, Y. (1958): The Determinants of Corporate Investment. Unpublished Ph.D. thesis (University of Chicago).

Halperin, M. (1951): "Normal Regression Theory in the Presence of Intraclass Correlation", Annals of Mathematical Statistics, 22, pp. 573-80.

Hartley, H. O. and J.N.K. Rao, "Maximum-Likelihood Estimation for the Mixed Analysis of Variance Model", Biometrika, 54, pp. 93-108.

Herbach, L. H. (1959): "Properties of Model II-type Analysis of Variance Tests", Annals of Mathematical Statistics, 30, pp. 939-59.

Hildreth, C. and J. P. Houck (1968): "Some Estimators for a Linear Model With Random Coefficients", Journal of the American Statistical Association, 63, pp. 584-95.

Hill, B. M. (1965): "Inference About Variance Components in the One-Way Model", Journal of the American Statistical Association, 60, pp. 806-25.

Hoch, I. (1962): "Estimation of Production Function Parameters Combining Time Series and Cross-Section Data", Econometrica, 30, pp. 34-53.

Houthakker, H. S. (1961): "An International Comparison of Personal Savings", Bulletin of the International Statistical Institute, 38, pp. 56-69.

__________ (1962): "On Some Determinants of Saving in Developed and Underdeveloped Countries", presented to International Economic Associations' International Congress on Economic Development, Vienna, 1962, and reprinted in E. A. G. Robinson (ed.), Problems of Economic Development, New York: MacMillan Company, 1965.

__________ (1965): "New Evidence on Demand Elasticities", Econometrica, 33, pp. 277-88.

Hurwicz, L. (1950): "Systems With Nonadditive Disturbances", Chapter 18 in T. C. Koopmans (editor), Statistical Inference in Dynamic Economic Models, New York: John Wiley & Sons.

Hussain, A. (1969): "A Mixed Model for Regressions", Biometrika, 56, pp. 327-36.

Jorgenson, D. W. (1965): "Anticipations and Investment Behavior", Chapter 2 of the Brookings Quarterly Econometric Model of the United States, edited by J. S. Duesenberry, G. Fromm, L. R. Klein and E. Kuh, Chicago: Rand McNally and Company.

Kakwani, N. C. (1967): "The Unbiasedness of Zellner's Seemingly Unrelated Regression Equations Estimators", Journal of the American Statistical Association, 62, pp. 141-42.

Klein, L. R. (1953): A Textbook of Econometrics, Evanston: Row Peterson and Company.

Kuh, E. (1959): "The Validity of Cross-Sectionally Estimated Behavior Equations in Time-Series Applications", Econometrica, 27, pp. 197-214.

________ (1963): Capital Stock Growth. A Micro-econometric Approach, Amsterdam: North-Holland Publishing Company.

Malinvaud, E. (1966): Statistical Methods of Econometrics, Chicago: Rand McNally and Company.

Marsaglia, G. (1965): "Ratios of Normal Variables and Ratios of Sums of Uniform Variables", Journal of the American Statistical Association, 60, pp. 193-204.

McHugh, R. B. and P. W. Mielke (1968): "Negative Variance Components and Statistical Dependence in Nested Sampling", Journal of the American Statistical Association, 63, pp. 1000-3.

Mehta, J. S. and P.A.V.B. Swamy (1970): "The Finite Sample Distribution of Theil's Mixed Regression Estimator and a Related Problem", Review of the International Statistical Institute, 38, pp. 202-9.

Mitra, S. K. and C. R. Rao (1968a): "Simultaneous Reduction of a Pair of Quadratic Forms", Sankhyā Series A, 29, pp. 313-32.

________ (1968b): "Some Results in Estimation and Tests of Linear Hypotheses Under the Gauss-Markoff Model", Sankhyā Series A, 30, pp. 281-90.

Mundlak, Y. (1961): "Empirical Production Functions Free of Management Bias", Journal of Farm Economics, 43, pp. 44-56.

________ (1963): "Estimation of Production and Behavioral Functions from a Combination of Cross-Section and Time Series Data", in C. F. Christ (editor), Measurement in Economics, Stanford: Stanford University Press.

Nelder, J. A. (1968): "Regression, Model-Building and Invariance", Journal of the Royal Statistical Society Series A, 131, pp. 303-15.

Nerlove, M. (1965): Estimation and Identification of Cobb-Douglas Production Functions, Chicago: Rand McNally and Company.

________ (1967): "Further Results on the Estimation of Dynamic Economic Relations from a Time Series of Cross-Sections", The Economic Studies Quarterly, , pp. 43-74.

Neudecker, H. (1969): "Some Theorems on Matrix Differentiation With Special Reference to Kronecker Matrix Products", Journal of the American Statistical Association, 64, pp. 953-63.

Parks, R. W. (1967): "Efficient Estimation of a System of Regression Equations When Disturbances are Both Serially and Contemporaneously Correlated", Journal of the American Statistical Association, 62, pp. 500-9.

Rao, C. R. (1963): "Criteria of Estimation in Large Samples", Sankhyā Series A, 25, pp. 189-200.

________ (1965a): Linear Statistical Inference and Its Applications, New York: John Wiley & Sons.

___________ (1965b): "The Theory of Least Squares When the Parameters are Stochastic and Its Application to the Analysis of Growth Curves", Biometrika, 52, pp. 447-58.

___________ (1966): "Generalized Inverse for Matrices and Its Application in Mathematical Statistics, pp. 263-80, in Research Papers in Statistics, F. N. David (editor), New York: John Wiley & Sons.

___________ (1967): "Least Squares Theory Using an Estimated Dispersion Matrix and Its Application to Measurement of Signals", pp. 355-71 in Proceedings of the Fifth Berkeley Symposium on Mathematical Statistics and Probability, Berkeley and Los Angeles: University of California Press, 1.

___________ (1968): "A Note on a Previous Lernma in the Theory of Least Squares and Some Further Results", Sankhyā Series A, 30, pp. 259-66.

___________ (1970): "Estimation of Heteroskedastic Variances in Linear Models", Journal of the American Statistical Association, 65, pp. 161-72.

Rosenberg, Barr (1970): "Varying Parameter Regression in the Analysis of a Cross Section of Time Series--I, Berkeley: University of California. Mimeo.

Rubin, H. (1950): "Note on Random Coefficients", Chapter XIX in T. C. Koopmans (editor): Statistical Inference in Dynamic Economic Models, New York: John Wiley & Sons.

Scheffé, H. (1959): The Analysis of Variance, New York: John Wiley & Sons.

Silvey, S. D. (1964): "A Note on Maximum Likelihood in the Case of Dependent Random Variables", Journal of the Royal Statistical Society Series B, 26, pp. 444-52.

Stone, M. and B.G.F. Springer (1965): "A Paradox Involving Quasi Prior Distributions", Biometrika, 52, pp. 623-7.

Swamy, P.A.V.B. (1968): Statistical Inference in a Random Coefficient Regression Model, Unpublished Ph.D. thesis (University of Wisconsin, Madison).

___________ and J. S. Mehta (1969): "On Theil's Mixed Regression Estimator", Journal of the American Statistical Association, 64, pp. 273-6.

Theil, H. (1954): Linear Aggregation of Economic Relations, Amsterdam: North-Holland Publishing Company.

___________ (1963): "On the Use of Incomplete Prior Information in Regression Analysis", Journal of the American Statistical Association, 58, pp. 1067-79.

___________ (1971): Principles of Econometrics, New York: John Wiley & Sons.

___________ and L.B.M. Mennes (1959): "Conception Stochastique de Coefficients Multiplicateurs dans L'adjustement lineaire des Series Temporelles", Publications de l'Institut de Statistuque de l'Universite de Paris, 8, pp. 211-27.

___________ and A. S. Goldberger (1961): "On Pure and Mixed Statistical Estimation in Economics", International Economic Review, 2, pp. 65-78.

Thompson, W. A., Jr. (1962): "The Problems of Negative Estimates of Variance Components", Annals of Mathematical Statistics, 33, pp. 273-89.

___________ (1963): "Nonnegative Estimates of Variance Components", Technometrics, 5, pp. 441-9.

Tiao, G. C. (1966): "Bayesian Comparison of Means of a Mixed Model With Application to Regression Analysis", Biometrika, 53, pp. 11-25.

__________ and A. Zellner (1964): "Bayes's Theorem and the Use of Prior Knowledge in Regression Analysis", Biometrika, 44, pp. 315-27.

__________ and W. Y. Tan (1965): "Bayesian Analysis of Random-Effect Models in the Analysis of Variance, I. Posterior Distribution of Variance-Components", Biometrika, 52, pp. 37-53.

__________ and __________ (1966): "Bayesian Analysis of Random-Effect Models in the Analysis of Variance, II. Effect of Autocorrelated Errors", Biometrika, 53, pp. 477-95.

__________ and N. R. Draper (1968): "Bayesian Analysis of Linear Models With Two Random Components With Special Reference to the Balanced Incomplete Block Design", Biometrika, 55, pp. 101-17.

Tracy, D. S. and P. S. Dwyer (1969): "Multivariate Maxima and Minima with Matrix Derivatives", Journal of the American Statistical Association, 64, pp. 1576-94.

Wald, A. (1947): "A Note on Regression Analysis", Annals of Mathematical Statistics, 18, pp. 586-9.

Wallace, T. D. and A. Hussain (1969): "The Use of Error Components Models in Combining Cross-Sections With Time Series Data", Econometrica, 37, pp. 55-72.

Watson, G. S. (1967): "Linear Least Squares Regression", Annals of Mathematical Statistics, 38, pp. 1679-99.

Williams, J. S. (1967): "The Variance of Weighted Regression Estimators", Journal of the American Statistical Association, 62, pp. 1290-301.

Zarembka, P. (1968): "Functional Form in the Demand for Money", Journal of the American Statistical Association, 63, pp. 502-11.

Zellner, A. (1962a): "An Efficient Method of Estimating Seemingly Unrelated Regressions and Tests for Aggregation Bias", Journal of the American Statistical Association, 57, pp. 348-68.

__________ (1962b): "Estimation of Cross-Section Relations: Analysis of a Common Specification Error", Metroeconomica, pp. 111-17.

__________, D. S. Huang and L. C. Chau (1965): "Further Analysis of the Short-Run Consumption Function With Emphasis on the Role of Liquid Assets", Econometrica, 33, pp. 571-81.

__________ (1965): "Bayesian Inference and Simultaneous Equation Econometric Models", Paper presented to the First World Congress of the Econometric Society, Rome.

__________ (1966): "On the Aggregation Problem: A New Approach to a Troublesome Problem", Report #6628, Center for Mathematical Studies in Business and Economics, University of Chicago. Published in Fox, K. A. et al. (editors): Economic Models, Estimation and Risk Programming: Essays in Honor of Gerhard Tintner, New York: Springer Verlag, 1969.

___________ and U. Sankar (1967): "On Errors in the Variables", Report #6703, Center for Mathematical Studies in Business and Economics, University of Chicago.

___________ and M. S. Geisel (1968): "Analysis of Distributed Lag Models With Application to Consumption Function Estimation", Manuscript, Center for Mathematical Studies in Business and Economics, University of Chicago. To appear in Econometrica.

Zyskind, G. (1967): "On Connonical Forms, Nonnegative Covariance Matrices and Best and Simple Least Squares Linear Estimator in Linear Models", Annals of Mathematical Statistics, 38, pp. 1092-110.

___________ (1969): "Parametric Augmentation and Error Structures Under Which Certain Simple Least Squares and Analysis of Variance Procedures are Also Best", Journal of the American Statistical Association, 64, pp. 1353-68.

Lecture Notes in Operations Research and Mathematical Systems

Vol. 1: H. Bühlmann, H. Loeffel, E. Nievergelt, Einführung in die Theorie und Praxis der Entscheidung bei Unsicherheit. 2. Auflage, IV, 125 Seiten 4°. 1969. DM 12,–

Vol. 2: U. N. Bhat, A Study of the Queueing Systems M/G/1 and GI/M/1. VIII, 78 pages. 4°. 1968. DM 8,80

Vol. 3: A. Strauss, An Introduction to Optimal Control Theory. VI, 153 pages. 4°. 1968. DM 14,–

Vol. 4: Einführung in die Methode Branch and Bound. Herausgegeben von F. Weinberg. VIII, 159 Seiten. 4°. 1968. DM 14,–

Vol. 5: L. Hyvärinen, Information Theory for Systems Engineers. VIII, 205 pages. 4°. 1968. DM 15,20

Vol. 6: H. P. Künzi, O. Müller, E. Nievergelt, Einführungskursus in die dynamische Programmierung. IV, 103 Seiten. 4°. 1968. DM 9,–

Vol. 7: W. Popp, Einführung in die Theorie der Lagerhaltung. VI, 173 Seiten. 4°. 1968. DM 14,80

Vol. 8: J. Teghem, J. Loris-Teghem, J. P. Lambotte, Modèles d'Attente M/G/1 et GI/M/1 à Arrivées et Services en Groupes. IV, 53 pages. 4°. 1969. DM 6,–

Vol. 9: E. Schultze, Einführung in die mathematischen Grundlagen der Informationstheorie. VI, 116 Seiten. 4°. 1969. DM 10,–

Vol. 10: D. Hochstädter, Stochastische Lagerhaltungsmodelle. VI, 269 Seiten. 4°. 1969. DM 18,–

Vol. 11/12: Mathematical Systems Theory and Economics. Edited by H. W. Kuhn and G. P. Szegö. VIII, IV, 486 pages. 4°. 1969. DM 34,–

Vol. 13: Heuristische Planungsmethoden. Herausgegeben von F. Weinberg und C. A. Zehnder. II, 93 Seiten. 4°. 1969. DM 8,–

Vol. 14: Computing Methods in Optimization Problems. Edited by A. V. Balakrishnan. V, 191 pages. 4°. 1969. DM 14,–

Vol. 15: Economic Models, Estimation and Risk Programming: Essays in Honor of Gerhard Tintner. Edited by K. A. Fox, G. V. L. Narasimham and J. K. Sengupta. VIII, 461 pages. 4°. 1969. DM 24,–

Vol. 16: H. P. Künzi und W. Oettli, Nichtlineare Optimierung: Neuere Verfahren, Bibliographie. IV, 180 Seiten. 4°. 1969. DM 12,–

Vol. 17: H. Bauer und K. Neumann, Berechnung optimaler Steuerungen, Maximumprinzip und dynamische Optimierung. VIII, 188 Seiten. 4°. 1969. DM 14,–

Vol. 18: M. Wolff, Optimale Instandhaltungspolitiken in einfachen Systemen. V, 143 Seiten. 4°. 1970. DM 12,–

Vol. 19: L. Hyvärinen, Mathematical Modeling for Industrial Processes. VI, 122 pages. 4°. 1970. DM 10,–

Vol. 20: G. Uebe, Optimale Fahrpläne. IX, 161 Seiten. 4°. 1970. DM 12,–

Vol. 21: Th. Liebling, Graphentheorie in Planungs- und Tourenproblemen am Beispiel des städtischen Straßendienstes. IX, 118 Seiten. 4°. 1970. DM 12,–

Vol. 22: W. Eichhorn, Theorie der homogenen Produktionsfunktion. VIII, 119 Seiten. 4°. 1970. DM 12,–

Vol. 23: A. Ghosal, Some Aspects of Queueing and Storage Systems. IV, 93 pages. 4°. 1970. DM 10,–

Vol. 24: Feichtinger, Lernprozesse in stochastischen Automaten. V, 66 Seiten. 4°. 1970. DM 6,–

Vol. 25: R. Henn und O. Opitz, Konsum- und Produktionstheorie I. II, 124 Seiten. 4°. 1970. DM 10,–

Vol. 26: D. Hochstädter und G. Uebe, Ökonometrische Methoden. XII, 250 Seiten. 4°. 1970. DM 18,–

Bitte wenden/Continued

Vol. 27: I. H. Mufti, Computational Methods in Optimal Control Problems. IV, 45 pages. 4°. 1970. DM 6,–

Vol. 28: Theoretical Approaches to Non-Numerical Problem Solving. Edited by R. B. Banerji and M. D. Mesarovic. VI, 466 pages. 4°. 1970. DM 24,–

Vol. 29: S. E. Elmaghraby, Some Network Models in Management Science. III, 177 pages. 4°. 1970. DM 16,–

Vol. 30: H. Noltemeier, Sensitivitätsanalyse bei diskreten linearen Optimierungsproblemen. VI, 102 Seiten. 4°. 1970. DM 10,–

Vol. 31: M. Kühlmeyer, Die nichtzentrale t-Verteilung. II, 106 Seiten. 4°. 1970. DM 10,–

Vol. 32: F. Bartholomes und G. Hotz, Homomorphismen und Reduktionen linearer Sprachen. XII, 143 Seiten. 4°. 1970. DM 14,–

Vol. 33: K. Hinderer, Foundations of Non-stationary Dynamic Programming with Discrete Time Parameter. VI, 160 pages. 4°. 1970. DM 16,–

Vol. 34: H. Störmer, Semi-Markoff-Prozesse mit endlich vielen Zuständen. Theorie und Anwendungen. VII, 128 Seiten. 4°. 1970. DM 12,–

Vol. 35: F. Ferschl, Markovketten. VI, 168 Seiten. 4°. 1970. DM 14,–

Vol. 36: M. P. J. Magill, On a General Economic Theory of Motion. VI, 95 pages. 4°. 1970. DM 10,–

Vol. 37: H. Müller-Merbach, On Round-Off Errors in Linear Programming. VI, 48 pages. 4°. 1970. DM 10,–

Vol. 38: Statistische Methoden I, herausgegeben von E. Walter. VIII, 338 Seiten. 4°. 1970. DM 22,–

Vol. 39: Statistische Methoden II, herausgegeben von E. Walter. IV, 155 Seiten. 4°. 1970. DM 14,–

Vol. 40: H. Drygas, The Coordinate-Free Approach to Gauss-Markov Estimation. VIII, 113 pages. 4°. 1970. DM 12,–

Vol. 41: U. Ueing, Zwei Lösungsmethoden für nichtkonvexe Programmierungsprobleme. IV, 92 Seiten. 4°. 1971. DM 16,–

Vol. 42: A.V. Balakrishnan, Introduction to Optimization Theory in a Hilbert Space. IV, 153 pages. 4°. 1971. DM 16,–

Vol. 43: J. A. Morales, Bayesian Full Information Structural Analysis. VI, 154 pages. 4°. 1971. DM 16,–

Vol. 44: G. Feichtinger, Stochastische Modelle demographischer Prozesse. XIII, 404 pages. 4°. 1971. DM 28,–

Vol. 45: K. Wendler, Hauptaustauschschritte (Principal Pivoting). II, 64 pages. 4°. 1971. DM 16,–

Vol. 46: C. Boucher, Leçons sur la théorie des automates mathématiques. VIII, 193 pages. 4°. 1971. DM 18,–

Vol. 47: H. A. Nour Eldin, Optimierung linearer Regelsysteme mit quadratischer Zielfunktion. VIII, 163 pages. 4°. 1971. DM 16,–

Vol. 48: M. Constam, Fortran für Anfänger. VI, 143 pages. 4°. 1971. DM 16,–

Vol. 49: Ch. Schneeweiß, Regelungstechnische stochastische Optimierungsverfahren. XI, 254 pages. 4°. 1971. DM 22,–

Vol. 50: Unternehmensforschung Heute – Übersichtsvorträge der Züricher Tagung von SVOR und DGU, September 1970. Herausgegeben von M. Beckmann. IV, 133 pages. 4°. 1971. DM 16,–

Vol. 51: Digitale Simulation. Herausgegeben von K. Bauknecht und W. Nef. IV, 207 pages. 4°. 1971. DM 18,–

Vol. 52: Invariant Imbedding. Proceedings of the Summer Workshop on Invariant Imbedding Held at the University of Southern California, June – August 1970. Edited by R. E. Bellman and E. D. Denman. IV, 148 pages. 4°. 1971. DM 16,–

Vol. 53: J. Rosenmüller, Kooperative Spiele und Märkte. IV, 152 pages. 4°. 1971. DM 16,–

Vol. 54: C. C. von Weizsäcker, Steady State Capital Theory. III, 102 pages. 4°. 1971. DM 16,–

Vol. 55: P. A. V. B. Swamy, Statistical Inference in Random Coefficient Regression Models. VIII, 209 pages. 4°. 1971. DM 20,–

Beschaffenheit der Manuskripte

Die Manuskripte werden photomechanisch vervielfältigt; sie müssen daher in sauberer Schreibmaschinenschrift geschrieben sein. Handschriftliche Formeln bitte nur mit schwarzer Tusche eintragen. Notwendige Korrekturen sind bei dem bereits geschriebenen Text entweder durch Überkleben des alten Textes vorzunehmen oder aber müssen die zu korrigierenden Stellen mit weißem Korrekturlack abgedeckt werden. Falls das Manuskript oder Teile desselben neu geschrieben werden müssen, ist der Verlag bereit, dem Autor bei Erscheinen seines Bandes einen angemessenen Betrag zu zahlen. Die Autoren erhalten 75 Freiexemplare.

Zur Erreichung eines möglichst optimalen Reproduktionsergebnisses ist es erwünscht, daß bei der vorgesehenen Verkleinerung der Manuskripte der Text auf einer Seite in der Breite möglichst 18 cm und in der Höhe 26,5 cm nicht überschreitet. Entsprechende Satzspiegelvordrucke werden vom Verlag gern auf Anforderung zur Verfügung gestellt.

Manuskripte, in englischer, deutscher oder französischer Sprache abgefaßt, nimmt Prof. Dr. M. Beckmann, Department of Economics, Brown University, Providence, Rhode Island 02912/USA oder Prof. Dr. H. P. Künzi, Institut für Operations Research und elektronische Datenverarbeitung der Universität Zürich, Sumatrastraße 30, 8006 Zürich entgegen.

Cette série a pour but de donner des informations rapides, de niveau élevé, sur des développements récents en économétrie mathématique et en recherche opérationnelle, aussi bien dans la recherche que dans l'enseignement supérieur. On prévoit de publier

1. des versions préliminaires de travaux originaux et de monographies
2. des cours spéciaux portant sur un domaine nouveau ou sur des aspects nouveaux de domaines classiques
3. des rapports de séminaires
4. des conférences faites à des congrès ou à des colloquiums

En outre il est prévu de publier dans cette série, si la demande le justifie, des rapports de séminaires et des cours multicopiés ailleurs mais déjà épuisés.

Dans l'intérêt d'une diffusion rapide, les contributions auront souvent un caractère provisoire; le cas échéant, les démonstrations ne seront données que dans les grandes lignes. Les travaux présentés pourront également paraître ailleurs. Une réserve suffisante d'exemplaires sera toujours disponible. En permettant aux personnes intéressées d'être informées plus rapidement, les éditeurs Springer espèrent, par cette série de »prépublications«, rendre d'appréciables services aux instituts de mathématiques. Les annonces dans les revues spécialisées, les inscriptions aux catalogues et les copyrights rendront plus facile aux bibliothèques la tâche de réunir une documentation complète.

Présentation des manuscrits

Les manuscrits, étant reproduits par procédé photomécanique, doivent être soigneusement dactylographiés. Il est recommandé d'écrire à l'encre de Chine noire les formules non dactylographiées. Les corrections nécessaires doivent être effectuées soit par collage du nouveau texte sur l'ancien soit en recouvrant les endroits à corriger par du verni correcteur blanc.

S'il s'avère nécessaire d'écrire de nouveau le manuscrit, soit complètement, soit en partie, la maison d'édition se déclare prête à verser à l'auteur, lors de la parution du volume, le montant des frais correspondants. Les auteurs reçoivent 75 exemplaires gratuits.

Pour obtenir une reproduction optimale il est désirable que le texte dactylographié sur une page ne dépasse pas 26,5 cm en hauteur et 18 cm en largeur. Sur demande la maison d'édition met à la disposition des auteurs du papier spécialement préparé.

Les manuscrits en anglais, allemand ou francais peuvent être adressés au Prof. Dr. M. Beckmann, Department of Economics, Brown University, Providence, Rhode Island 02912/USA ou au Prof. Dr. H.P. Künzi, Institut für Operations Research und elektronische Datenverarbeitung der Universität Zürich, Sumatrastraße 30, 8006 Zürich.

ISBN 3-540-05603-3
ISBN 0-387-05603-3